Absolute Relativity: How Newton and Einstein Agree

By

Timothy Michaels

First Edition v1.1

Cover Photo by Gerd Altmann

Table of Contents

Preface

This book presents a new theory of relativity to both the scientific community and to casual readers of science. I've included the arguments that scientists may need in order to evaluate it, as well as explanations which I believe will help casual readers and non-scientists understand.

In each section, there are some particulars that may be difficult to understand. However, I think that I've provided a good explanation of enough of the fundamentals that most readers will be able to follow all the ideas, and all readers will be able to follow most of the ideas, well enough to learn what absolute relativity is about.

Beyond presenting a new idea, I hope I can contribute to each reader's understanding of physics, and their ability to evaluate the world around them.

The idea for absolute relativity came out of the process of fitting together relativity with the quantum theories. The assembly of the two physics is the subject of my book, "*Out of This World: The Movement Dimension.*" While the concept of the movement dimension may make sense to a casual reader, someone familiar enough with the current understanding of relativity may notice that I did something new which I didn't explain in the book.

What I did was to apply absolute space and time (as Isaac Newton believed in it). To make that work, I made space and time dilation into "experience" of space and time, rather than the actual changes of space and time, which are currently believed to occur.

So, in this book I will explain the version of relativity which I've applied. It lays out the relativity principles which support the movement dimension concept, and which allow relativity and quantum theories to fit together.

It's the special and general theories of relativity which are incompatible with quantum theory.

Absolute relativity, as I describe here, invokes Newtonian ideas to form a version of relativity which fits with the quantum theories, tells us more about our world, and I think makes more sense. I hope you'll agree. Or, at least that you may be entertained and provoked to give it some thought.

Chapter 1 - Introduction

Our laws of physics require a continuum to operate from. This need is derived from relativity. We need a smooth (continuous) arena for our mechanical world to function in order for physics to be reliable. But ours is pixelated, not smooth.

The quantum theories are based on the quantization of space, time, energy, and therefore matter. This conflict indicates that there must be someplace else. (I use the term "place" loosely here.)

Not only can physics <u>not</u> function in quantized space and time, but matter can't even move there. Movement can't occur in quantized space and time. However, since movement <u>does</u> occur, it must occur someplace else.

One of the mysteries Einstein's relativity brought us was the loss of time and distance experienced by everything which moves. The rate of loss is dependent on speed.

I believe that there is another dimension that exists to accommodate movement, a place where everything goes during travel. In my previous book "*Out of This World: The Movement Dimension*", I explain that this is not just a way of understanding how movement occurs and how relativity and quantum theories work together, but it's the true behavior of our world. It is predictable and measurable according to Lorentz's dilation formulas, which tell us how much distance and time is lost by a traveling object.

To prove that objects actually leave and come back repeatedly as they travel, we must watch one. The difficulty is that an object slow enough for us to see is leaving space for such a small amount of time that we can't detect it. How could we know it was gone for a billionth of a trillionth of a trillionth of a second?

Objects fast enough to leave space noticeably, are too fast to see. But there is a sign of this leaving and returning, and that is that solid objects will pass through each other. Two solid objects, each traveling fast enough will pass through each other without colliding.

The reason this will occur is the wave/particle duality of all things. Everything is a wave, and also a particle. Popular understanding is that matter is particles which have wave-like properties but are not actually waves, and energy is waves which have particle-like properties but are not actually particles. My proposed movement theory tells us that this is not true. Instead, all things alternate between being matter and being energy. This is how movement occurs.

Matter cannot move. Waves must move, and they must move at full speed, the speed of light. The speed of an object tells us how much an object is a wave and how much it is a particle. And this proportion can be calculated with the dilation formula.

Whether an object is a particle or a wave is very important when it comes to collisions. In fact, collision is one of the main indicators of which type of object it is. Waves cannot collide with waves. They pass through each other. Particles cannot pass through each other. They will always collide.

Movement theory tells us that objects that are not moving are particles and cannot be passed through. Objects moving at the speed of light are waves and will pass through each other. And all else, at speeds in-between, have the ability to pass through or collide depending on whether they're a particle or wave when they encounter each other. The chances of which action will occur can be determined using the dilation formula.

This may seem incorrect as we look around and see that all matter always collides. It's only pure energy, such as radio, television, cell phone, light, heat, and other radiation which passes through each other. But we must consider how extremely low the probability is of everyday objects passing through each other because of their high number of particles and very low speed.

If you and I ran toward each other, how fast would we have to go in order to not collide? The average person is estimated to be made up of 10^{28} atoms. Every one of yours must pass through every one of mine, lest we collide or get stuck together as a single body in our attempt. Every particle would have to be in wave form

4

simultaneously for us to pass through. If we only wanted to pass 1 atom through 1 atom, both must go 70.7% of light speed (474,000,000 mph) in order to have a 50% chance of success. There is a 50% chance that both atoms will be in wave form at the moment they encounter each other. For a greater chance, they would have to go much faster.

For 2 atoms to have a 50% chance of passing through 2 atoms, both pairs must go 84.1% of light speed (564,000,000 mph). For you and I to have a 50% chance of passing all of our 10^{28} atoms through each other we would each need to travel 99.9999999999999999999999993% of light speed (over 670,000,000 mph). This is the reason why we don't find solid objects passing through each other. It is far too improbable to occur in our world.

But there is an environment in which this <u>can</u> happen. And, it is controlled in such a way that we already have the data we need to investigate: particle accelerators. These devices collide particles at the high speeds which would result in a measurable degree of them passing through each other. Later we'll take a look at some data which reveals that it may be happening. And, this is evidence that movement does occur according to the movement dimension concept. But for now, on to the theory of relativity which is behind it all.

Chapter 2 – General vs Absolute Relativity

The difference between the general theory of relativity (which includes the special theory) and absolute relativity is that absolute relativity acknowledges a preferred observer, and the general theory does not. Absolute relativity recognizes that there is such a thing as absolute rest. And this solves the paradox of the multiple realities which arise from the general theory of relativity.

In 1905, in his paper "On the Electrodynamics of Moving Bodies", Einstein said, "...the phenomena of electrodynamics as well as of mechanics possesses no properties corresponding to the idea of absolute rest." He points out that electric current is created in a wire whether a magnet is moving past the wire, or the wire is moving past the magnet. This, to him, indicated that there is no state of rest. Our physical laws work the same for any uniformly moving reference frame. They do not account for the existence of absolute rest. They dictate individual relationships only.

In 1908, Hermann Minkowski, in his paper "Space and Time", built on that idea by introducing his axiom of "The substance at any world-point may always, with the appropriate determination of space and time, be looked upon as at rest."

It's popularly believed that because many laws of physics don't acknowledge a state of absolute rest, that any uniformly accelerating state may be considered as being at rest. And this is how we obtain reference points for making predictions. We choose an observer whose experience we're interested in.

However, I don't believe that just because many of the laws don't acknowledge absolute rest means there isn't any. I believe all the laws rely on absolute rest to operate from. And, that this state is revealed to us by the behavior of the fundamental substance of nature, the one thing which moves only according to absolute rest: energy.

This is what I intend to show in the coming pages. I believe energy (light) reveals to us the state of absolute rest. So, I'll be showing you the evidence as well as how it may help us to learn more about our world.

Evolution of Relativity

The term "relativity" is recent, though the concept is not. It only became popular because of Einstein's revisions to it in 1905 and 1915. Before then it was so obvious and foundational most people never thought about it and so they didn't call it anything. That's why when Einstein discovered how consequential it could be, he needed to call it something. He wanted to call it his "invariance theory", but Henri Poincare' insisted it be called "relativity".

Relativity is a fundamental concept which science is built on. It establishes that there is sameness and identifies what the variables are which affect how our world works. It is simple and can be expressed in one sentence. Here is that one sentence as it has evolved.

"The world is knowable." This seems to be the first version of relativity. It dates back at least to Thales of Miletus, around 600 BC. From this concept we could begin to identify causes to events and make predictions. This was a realization that there are physical laws which are unchanging regardless of place or time.

"The laws of motion are the same for any uniformly moving reference frame." This was relativity according to Galileo Galilei (1564-1642). He believed that the natural state of an object or reference frame, the location where you're at, is to be in motion. So, his concern was not whether there was motion on the part of the observer, but whether it was smooth enough motion that he felt like he was still. The feeling of stillness allows all movement within that reference frame to occur the same as usual, the same as in any other smoothly moving reference frame.

Isaac Newton (1642-1727) agreed with Galileo that the natural state of motion was uniform motion. And, he believed this motion was related to something, space and time. He wrote,

"Absolute space, in its own nature, without relation to anything external, remains always similar and immovable,..." And, "Absolute, true and mathematical time, of itself, and from its own nature, flows equably without relation to anything external."

"*The laws of 'physics' are the same for any uniformly moving reference frame.*" In 1905, Albert Einstein (1879-1955) expanded relativity to say that not just the laws of "motion" work the same, but the laws of "physics" work the same in any uniformly moving reference frame. This covers all events, all changes in our world.

In Galileo's day all physics was thought to be motion, but by Einstein's time we had learned that some events are not based on spatial changes. These include electrical and magnetic activities. This was his special theory of relativity, and he did this specifically to explain how light moves at the same speed for every observer.

He went against Newton's belief in space and time being absolute and replaced it with an absolute speed of light. It revealed what we now refer to as "relativistic effects", which are the adjustments of distance and time to accommodate this constant speed of light. Simply stated, the faster an object is moving, the more time and space are reduced for that object so that the speed of light can remain the same for that object.

In order to explain how this is possible, Einstein described space and time as warping to accommodate these relativistic effects. The concept is believed to have been inspired by Bernhard Riemann (1826-1866) who suggested that the universe may be curved into a ball. He had expanded on the curved geometry concept which Carl Friedrich Gauss (1777-1855) had begun. And also William Clifford (1845-1879) who described the possibility of hills in small portions of space.

From learning that all movement is related through light, came the equivalency of matter and energy which we summarize with the equation $E = mc^2$.

"The laws of physics are the same for any uniformly 'accelerating' reference frame." In 1907, Einstein explained that gravity and acceleration are equivalent. By 1915, he had integrated this into relativity. This change doesn't alter the predictions of his special theory of relativity. It expands relativity to encompass more situations. Non-moving and uniformly moving reference frames are still covered because they have a uniform acceleration of zero. And now, in addition to them, all constant non-zero acceleration is included, such as an airplane or rocket while increasing its speed steadily. The effect is the same as increased gravity or a change in its direction, so all laws of physics, then, work the same.

This led to the relativistic effects, time and distance dilation, being predicted as a result of gravity in addition to movement.

Proposed Update to Relativity

"The laws of physics are the same for any uniformly accelerating reference frame, 'through space and time'." This is my proposed update to relativity. It doesn't broaden the reference frames which laws of physics may be applied to. It specifies what the laws of physics are relative through. It establishes that spacetime is the reference frame where the laws of physics operate from so that they may be the same in all reference frames which are uniformly accelerating through it. I bring back Newton's absolute space and time.

This new version doesn't change any of the fundamental workings of the general theory of relativity. It allows us to take its predictions further and add a couple new ones. So, we'll look at Einstein's relativity first. Then we'll explore how having a common reference frame to draw from can allow us to make further predictions.

Chapter 3 – The General Theory of Relativity

From Einstein's update to relativity, we've been able to advance our understanding of the world based on a few major concepts.

1. The speed of light is constant, and time and space adjust themselves to accommodate that. This variability, or dilation, since it is a reduction of distance and time, is referred to as the relativistic effects. And because the energy of light is dependent on distance and time, we get energy dilation also.
2. Mass and energy are equivalent through the speed of light.
3. Gravity creates effects the same as acceleration. It causes the same dilation of distance and time which movement does.

The Speed of Light Is Constant

A photon doesn't exist until it's emitted by matter. A photon is a piece, a quantum, of pure energy. It's called electromagnetic because it's made of alternating electric and magnetic fields. Each field generates the next. In this way there isn't so much a "thing" moving as a progression of fields. Fields which create each other. This is how we know that light moves at a constant rate. It moves itself according to physical principles which James Clerk Maxwell (1831-1879) laid out for us. Nothing else moves it.

We can depict these alternating fields as sine waves, each transitioning back and forth along 2 directions, one up-and-down, and one side-to-side, to show how each leads to the other.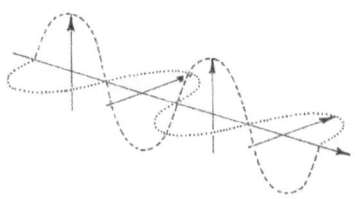

Maxwell's explanation is that the speed at which energy moves is equal to the ratio of 2 electric charges at rest compared to the magnetic force they create when in motion.

Whether moving toward or away from a light source, its speed compared to the observer is always the same. Maxwell's equations (physics) apply in all uniformly moving reference frames. This sameness fits with the idea that the world is knowable. But it creates a difference at the same time. The world is knowable to each person individually.

If I throw a ball to you at 10 mph while I run toward you at 10 mph, then to me the ball is going 10 mph, but to you it's going 20 mph. We both understand that, to you, the ball is moving faster than it is to me, because we understand we have different relationships with the ball. Mine is as a thrower, regardless of my own speed. Your relationship with the ball is as receiver of a ball being thrown 10 mph from a thrower running toward you at 10 mph. Your relationship with the ball is through your relationship with me. So, you expect a 20 mph ball. It's normal for us to experience different speeds of the same ball while we play the same game. We understand our different experiences of the speed of the ball.

If, somehow, we experienced the same speed of the ball, then it would seem like we're not playing the same game. Because to accomplish that I would experience my normal running and throwing while you would see me running at half my speed and throwing at half the speed so you could receive it at the same speed I throw it at. By coordinating one aspect of our experience (the speed of the ball), we distort the others.

This is the situation we run into with light. But because light travels so fast its effects are very small in most of our situations.

Instead of a ball, if I run toward you with a flashlight, you would receive the light at the same speed I'm sending it, not 10 mph faster as we might expect. As I run to you, my perception is that you're coming toward me at 10 mph and my light is going toward you at 670,616,629 mph. So, to me, the light should hit you at light speed plus 10 mph. But to you, I'm coming at you at 10 mph and the light is only coming to you at light speed (670,616,629 mph). You and I could measure the same event and get different results if we were able to notice that tiny fraction difference of 10 / 670,616,629. This seems to suggest that our realities are very slightly different.

Relativity tells us that the faster we move, the less time and distance seem to be in our direction of movement. The results are all uniquely based on the observer. This is because the speed of light is the same for every observer, but distance and time are not.

Dilation In Direction of Movement

The popular interpretation of relativistic effects is that time and distance in the direction of movement get reduced. Since movement is in relation to the observer, it's the bystander who would notice this, not the traveler. The traveler would see himself as not moving, and view the bystander as the one moving and experiencing the dilation (reduction).

Watching a very fast rocket fly by at 500,000,000 mph should cause a bystander to see a short fat rocket going 500,000,000 mph.

A person in the rocket would see the rocket as being at rest. So, they would see the Earth fly by at 500,000,000 mph. And the Earth wouldn't look round. It would appear oval. It would appear squashed by 33% due to the Earth's rate of movement relative to the observer in the rocket.

In 1892, Hendrik Lorentz wrote equations to explain how objects shrink in the direction of their movement. And when Einstein wrote his special theory of relativity in 1905, he referenced Lorentz's explanation for distance and time dilation to describe how light can maintain a constant speed for all observers.

The equations are:

$$T_1 = T_2\sqrt{1 - (v/c)^2}$$

T_1 = Time measured by traveler

T_2 = Time measured by bystander

v/c = Speed as a portion of light speed

$$D_1 = D_2\sqrt{1 - (v/c)^2}$$

D_1 = Distance measured by traveler

D_2 = Distance measured by bystander

v/c = Speed as a portion of light speed

Distance and time dilate equally in every situation.

The simple ratio "v/c" in which "v" represents the object's speed and "c" represents light speed, is commonly used in relativity since many of the interesting effects are a result of how fast something is moving as a portion of light speed.

In this terminology, you could say that when you're driving on the highway at 67 mph you're going 0.0000001c.

(67 mph / 670,000,000 mph = 1/10,000,000 of light speed)

Using these formula's we can see that the relativistic effects, time and space dilation, become dramatic as an object approaches light speed.

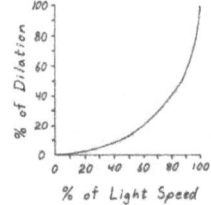

When both observers are moving, we use:

$$v_2 = \frac{u + v_1}{1 + uv_1}$$

v_2 = the total speed of the second object

v_1 = the speed of the second object compared to the first

u = the speed of the first object

With this we can see that if you're going 75% of the speed of light (0.75c), and you measure my speed as 0.75c faster than you, then my true speed is 96% of light speed (0.96c).

If you're going 0.9c and shine a flashlight ahead of you, you'll see the light traveling at 1c, and it will actually be traveling 1c.

We can see how every observer, regardless of his speed, always measures light speed as being the same, since both time and distance are reduced equally.

How Dilation Occurs

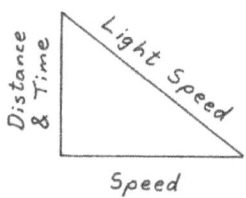

Time and distance adjustments occur according to the Pythagorean theorem in which the 3 sides of the triangle are speed, distance and time factor, and the speed of light. While we typically know the two legs of the square angle and are looking to find the hypotenuse (diagonal), in this case the hypotenuse is the speed of light and we're looking for how the legs of speed and distance and time factor work together to keep the speed of light constant.

Our dilation formula: $a = \sqrt{1 - (v^2/c^2)}$ is a version of the Pythagorean theorem. The form we're familiar with is $a^2 + b^2 = c^2$, in which "c" is the hypotenuse and "a" and "b" represent the other two sides.

This is how it translates. Let's say "a" is our distance and time factor, and "b" is speed, since "c" must be the speed of light. Our distance and time factor is expressed as a portion of full distance and time (as it would be at rest). So "1" is full distance and time, and "0" is no distance and time (fully dilated.) Speed is expressed as a portion of light speed. This means "b" is the " v/c" which we have been using. If our speed is half of light speed, then "b" would be 0.5.

The speed of light being "c", must also be expressed as a portion of light speed. So, "c" is always "1", or full light speed. To express it as a portion, we'll write it as (c/c).

15

The Pythagorean theorem:

$$a^2 + b^2 = c^2$$

then becomes $a^2 + (v/c)^2 = (c/c)^2$.

If we know our speed and want to solve for distance and time factor, we can move (v/c) to the other side of the equation:

$$a^2 = (c/c)^2 - (v/c)^2$$

To get "a" without the exponent, we take the square root of both sides and the result is:

$$a = \sqrt{(c/c)^2 - (v/c)^2}$$

We know that "c/c" is 1, and 1^2 is 1. This could have been simplified much earlier, but I wanted to show what the 1 is all about. The 1 we see in the dilation formula is to determine how much less than the speed of light squared, speed squared is. So, a good way to see how the formula works is to express it as

$$a = \sqrt{(c^2 - v^2)/c^2}$$

For ease of use, we reduce it to:

$$a = \sqrt{1 - v^2/c^2}$$

Having seen the transformation from Pythagorean theorem to dilation formula, it's easy to see that the distance and time factor and speed are interchangeable. We could have just as easily arrived at:

$$v/c = \sqrt{1 - a^2}$$

And this tells us the dilation curve is symmetrical. In fact, it's also a constant curve, a quarter of a circle with a radius of light speed as a portion of light speed, c/c or 1.

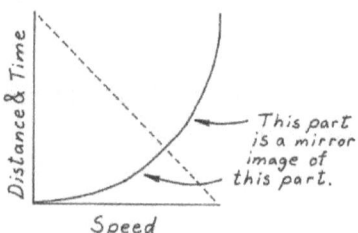

16

Viewed in triangular form, we can see that more plainly. At low speeds our distance and time factor is great. At low distance and time factor our speed is great.

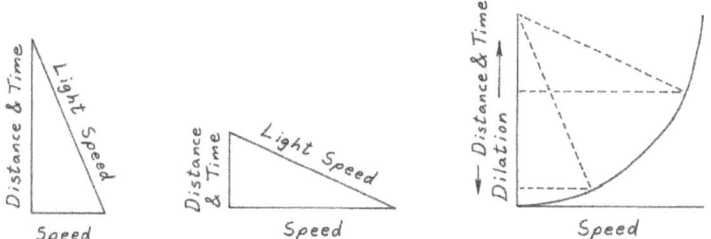

Note that distance and time factor and speed move opposite each other. They are reverse ways of looking at the same phenomenon.

Energy Dilation

We can apply dilation of distance and time to energy level of radiation (light) and find that energy dilates too.

Energy, electromagnetic radiation, which includes light, X-rays, microwaves, ultra-violet and infrared, radio, and television signals, and more, comes in the form of waves. Since energy always travels at the speed of light, and distance and time can be reduced, then the measurement of our wave's length gets reduced, and frequency gets increased. The same photon (packet of energy) with increased frequency (reduced time) and reduced wavelength (distance) has a higher energy level for the observer who's experiencing relativistic effects.

The observer who's moving, experiencing reduced distance and time, experiences light at a higher energy than an observer who's not moving.

Matter and Energy Are Equivalent

Soon after publishing his special theory of relativity, Einstein realized that energy and matter must be related through

17

light. As he wrote in a letter to Conrad Habicht, "The principle of relativity, in conjunction with Maxwell's equations, requires that mass be a direct measure of the energy contained in a body; light carries mass with it."

He proposed, "The mass of a body is a measure of its energy content; if the energy changes by "L", the mass changes in the same sense by $L/9 \times 10^{20}$ if the energy is measured in ergs and the mass in grams." He used "L" to represent energy, and 9×10^{20} is the speed of light in centimeters squared.

Today we use "c" to represent the speed of light and "E" to represent energy. So, a change in mass is equal to E/c^2, or $m = E/c^2$. This was his first version of the famous equation. The popular preference is to multiply both sides of this equation by c^2 to get $mc^2 = E$, or $E = mc^2$. It's the same equation in a different form, as he originally intended to solve for the mass of radioactive radium being lost due to radiation given off.

The reason why Einstein's updated (special) theory of relativity must equate matter and energy has to do with inertia. Since inertia of mass can do work just like energy can, then there must be some equivalency between them.

You may imagine the equivalency like this: You need to lift a box to a certain height. You could set up a pulley system and attach a weight to do the lifting. In this way the falling weight does the work.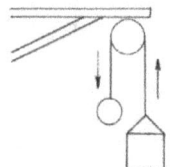

Or, you could attach an electric motor to wind up the rope and lift the box at the same speed. The electricity you use is equal to the amount of work accomplished by the hanging weight falling at that speed (not considering energy lost to friction), the potential energy the elevated weight had.

Einstein's realization that matter and energy are equivalent came by combining Maxwell's explanation of light, with relativity. But, he used the dilation effect, which resulted from the combination to come to that equivalency.

Since distance and time are dilated by speed, then wavelength and frequency are, too. Wavelength, which is a distance, becomes shorter. Frequency, the time period of waves, also becomes shorter, making frequency go up. This accommodates the constant speed of light.

From Planck's work in 1900, we know that energy and frequency are directly related. We see that in his formula of $E = hf$ (*Energy* = *Planck's constant x frequency*), which is how we know energy dilates, too.

Einstein applied dilation to the kinetic energy of an electron to explain why when an electron orbiting an atom emits a photon it simultaneously slows. Electrons lose speed when they emit photons. They lose kinetic energy when they emit radiation energy.

He, then, considered Newton's second law, which equates force with mass x acceleration ($F=ma$). This is a very interesting equation because it does in fact equate matter and energy. And, it translates directly into $E=mc^2$.

This is how. Force is energy divided by distance. Acceleration is speed divided by time, which is actually distance divided by time divided by time ($d/t/t$).

So we can expand $F=ma$ to be

$$E/d = md/t/t$$

or

$$E/d = md/t^2$$

We can multiply both sides by "d" to get

$$E = md^2/t^2$$

Since distance over time is speed, we can say that

$$E = mv^2$$

If light speed is used, this formula becomes

$$E = mc^2$$

19

One might be concerned that this formula isn't exactly correct because it recognizes any speed as being valid and doesn't acknowledge the cosmic speed limit of light speed. It was written about 200 years before we knew there was a speed limit.

But even if Einstein recognized that full translation, he didn't just accept its conclusion. He went on to consider that if an electron could emit a piece of its kinetic energy as a photon, then what is radium emitting a piece of when it puts out gamma rays (high energy photons)? If the radium isn't moving, then it must be emitting a piece of its mass. He realized that matter is made of energy. And he decided that the amount of mass radium must be losing is the energy emitted, divided by the speed of light squared. Therefore, $m = E/c^2$, which is $E = mc^2$.

In order to best understand this discovery we should look at the formula as $m = E/c^2$. This way it tells us what mass is. Energy is the fundamental substance which matter is composed of.

In Einstein's words from 1916, "The special theory of relativity has led to the conclusion that inert mass is nothing more or less than energy, ..."

This doesn't mean that $E=mv^2$, which is derived from Newton's $F=ma$, is reliable at all speeds. Speed creates dilation, and $F=ma$ does not acknowledge that. So if we want to be precise, then $F=ma$ may be modified to be:

$$F = \frac{mv^2}{d} \cdot \frac{1}{\sqrt{1 - v^2/c^2}}$$

(This is my modification, but I believe it's a valid relativistic version of Newton's original since force should become infinite at light speed.)

If a mass at rest could transition directly into energy of motion (kinetic energy), gradually and completely, it would become light. Along the way, as a moving body, it's transition would look like this:

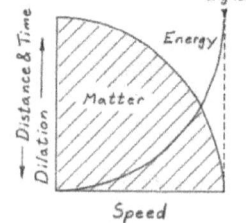

20

I believe this can be calculated using Einstein's formula for kinetic energy in combination with $E=mc^2$, arranged to solve for mass.

$$m = \frac{E}{c^2} \sqrt{1 - v^2/c^2}$$

m = mass in kilograms

E = total energy in Joules

v = speed in kilometers / second

c = the speed of light as 299,792,458 m/s

Total energy (which includes the energy equivalency of matter) remains constant in this hypothetical exercise, while mass decreases and speed increases. We are exchanging mass for speed.

I've also superimposed the dilation effect onto the chart to show how an object's proportion of mass to kinetic energy relates to dilation. (For this I assumed our object has no significant gravity which would contribute to dilation.)

Movement brings matter closer to becoming light and having no experience of time or distance.

If you'd like to experiment with $E = mc^2$, use these units:

(E) energy in Joules

(m) mass in kilograms

(c) speed of light as 299,792.458 m/s

or

(E) energy in ergs

(m) mass in grams

(c) speed of light as 29,979,245,800 cm/s

(1 food Calorie = 4190 Joules = 41,900,000,000 ergs)

Equivalence of Gravity and Acceleration

Just as speed reduces time and distance, gravity does also. It does this by drawing everything toward massive objects (objects with mass). We're familiar with gravity attracting mass. It also attracts energy, which we know is equivalent to mass.

Matter and energy are always being pulled toward a gravitational source, which is any other matter or energy. Because of this, every time an object (matter or energy) moves, its time and distance measure is reduced.

On Earth, to move upward, you have to lift your weight. This is much harder than walking on level ground. And, to move down is easier.

Likewise, if you shine light parallel to Earth's surface, it will have its normal wavelength and wave frequency. (It will also have its normal energy.)

Shine the same beam of light upward though, and the light still travels at the speed of light, but the waves will change because the waves are where the mass equivalency is. The waves (energy) struggle to go up, just like you do. So, the waves also go up slower (less frequently). Since the speed of light is always  the same, the length of the waves increase to accommodate the reduced speed of the waves. The constant speed of light stretches its slower moving waves. Therefore, light moving away from a gravity source has a reduced frequency (stretched wavelength), with a lower energy than usual.

If you shine a light down (toward a gravity source) the waves are attracted to it and have an easier time traveling. The waves occur more frequently and they're shorter because the speed of light remains the same. They have higher energy. This effect causes time and distance to be reduced due to gravitational pull.

The amount of this effect depends on gravity. And we can calculate it as if it were speed. The speed of falling.

Newton found that gravity can be calculated as

$$Gm/R^2$$

G = the gravitational constant of 6.672 x 10^{-11} Nm²/kg²

m = the mass of the gravitational body in kilograms

R = the distance from the center of gravity in meters

Since the Earth is 6,378,000 meters in radius, light we encounter here on the surface is 6,378,000 meters from the center of gravity. And Earth's mass is 5.974 x 10^{24} kilograms.

Using Newton's formula, we find that Earth's gravity is 9.8 m/s². Since a Newton of force accelerates 1 kg at 1 m/s², we are able to cancel units within the equation to leave us with only m/s².

$$\left(\frac{kgm^3}{kg^2s^2}\right) \cdot \left(\frac{kg}{m^2}\right) = m/s^2$$

Matter is doing work by pulling other matter closer, and in the process is slowly converting to energy in the form of gravitational waves. Gravity is a form of decay of matter (and energy).

Gravitational Decay Proven

In 1974, Joe Taylor and Russell Hulse discovered a pulsar (now known as PSR 1913+16) with a radio telescope (a device which detects energy well below our visible frequencies). They found the pulsar to be spinning 17 times per second and varying every 8 hours, which means it is a binary system. It is 2 neutron stars orbiting each other very closely. The general theory of relativity predicts the orbit to slow as energy is lost in the form of gravitational waves. For this system, the prediction was a reduction of orbit by 77 microseconds each year. They measured and recorded the pulsar's orbit speed for years. Their data matched predictions, and in 1993 they were awarded the Nobel Prize for their work.

Gravitational Dilation Proven

In 1959, at Harvard University, Robert Pound and Glen Rebka measured the difference in the apparent rate of time at ground level compared to how fast time passes at the top of a 22.5 meter tall tower.

They used gamma rays coming from radioactive Iron-57 because of its precise energy emission (3.5×10^{18} Hertz) and found that those gamma rays changed their frequency by 3×10^{-15}. From atop the tower the energy received from the iron on the ground was 3×10^{-15} less.

This is predicted by general relativity which explains that wave emissions at greater gravity (closer to Earth) are slowed by gravitational pull.

Had the gamma rays been measured at ground level they would have appeared normal because the measuring device would be running slower, also, and experiencing the same dilation.

Effects of Speed and Gravity On Time

In 1971, Joseph Hafele and Richard Keating flew atomic clocks around the world. They went both east and west. Because they flew "coach" on commercial flights, each trip took a few days and multiple flights. Before and after traveling, they compared their cesium clocks to a matching clock at the U.S. Naval Observatory in Washington, D.C.

- The trip east showed a total loss of 59 nanoseconds.
- The trip west showed a total gain of 273 nanoseconds.

These time differences are a result of both time dilation from speed of travel as well as gravitational dilation. Traveling fast slows the clock, while the high altitude helps it run faster. The clock on Earth has less gravitational force to slow it.

What really made the difference, though, was the rotational movement on Earth, which is faster than the aircraft speed. Traveling east, the total speed was aircraft speed plus Earth's rotation speed. Going west it was the difference between the two. Flying west is like undoing the rotation of the Earth which is why the gain from less gravity dominated that result.

Our Mechanical World Experience

Our experience of the world is mainly physical. It's very much about distances and times. While there are non-mechanical aspects to it, such as electrical and magnetic forces and various other properties of particles, what we experience is usually the mechanical results of them. So, most everything in our world converts to movement, and movement is what we experience. The world to us is mechanical, as we believed until the 1800's, but with underlying non-mechanical parts.

Being mechanical, it's constantly affected by relativity. Everything that moves, which many people believe is everything, is constantly experiencing adjustments to distance and time.

In the popular view of these relativistic effects, distance (space itself) and time are constantly warping to accommodate those adjustments. Like when a dog walks across a well-made bed, the fabric of our world is constantly adjusting. In this view, only light speed is absolute.

Most of us accept this explanation of constantly varying reality, and the unique realities for each person. This is because the variations are so slight that we aren't able to notice them in our daily lives. One could say it's only of concern to scientists. But that doesn't mean it's not a problem; it only helps us to not think about it.

I see it as a very small problem which has some big opportunities associated with it if it can be solved. The opportunities come by way of assembling these various adjustments to time and distance into a greater knowledge of our world.

And, I think the solution is a simple update to relativity involving the recognition that there is absolute movement.

Chapter 4 - Absolute Relativity

Absolute relativity is an interpretation of relativity in which there is one reality. All movement is relative to each observer through a common absolute movement.

In a theory of relativity in which every observer may have his own reality, it may be said that the speed of light is based solely on the observer.

But, in a theory of absolute movement, everything, including light, needs to have absolute movement. And, since we know that light speed measures the same for every observer, there must be a device which causes this to occur out of absolute movement.

It's my belief that the flexibility of space and time is not in actual distance and time changes, but in each observer's experience of the same distance and time.

In this way both Newton and Einstein are correct. Newton is correct in his belief that space and time are absolute. And Einstein is right with the speed of light being absolute and space and time adjusting to accommodate that. Only the adjustment of space and time is in the form of experience of them, not actual changes in space and time. I describe the way experience is able to vary in my book *"Out of This World: The Movement Dimension"* in which I apply relativity and quantum theories to the problem. Here, I'll be explaining how we can be sure that movement is absolute and how to apply this absolute concept in our world.

This view eliminates the paradox of different observers having different realities. It also tells us more about the world we live in. It tells us of deep space and about our place in the universe.

We know that movement is relative between objects which have some type of interaction. But what about objects that don't have any interaction? Is there any relationship between those objects? If there's no relationship, then when objects separate, they exist independent of each other. They might come back together having been apart by distances and for times which they can't agree on. This is how relativity seems to be commonly understood, and it's the problem I will address.

I believe the predictions of relativity only tell us of our differences of experience, not of actual differences. So, if 2 objects separate and have no known relationship, then come back together having had different experiences of time and distance, how is it that they were imperceptibly related? When there is nothing else, there is space and time.

It is the grid of space and time which true movement is relative to, and which we have our own varied experience of.

I'll go through two explanations of how we know that movement is absolute and it's only the experience which changes. After each proof, I'll point out some ways we can apply the effects of relativity to learn more about our world.

But before considering how we can discover absolute movement, lets first establish what this structure is which I intend to rely on, and which I will define absolute movement by.

Chapter 5 - Spacetime

The combined "spacetime" as being the physical world we live in, and even the concept of time as a dimension, come to us from Hermann Minkowski (1864-1909). In 1908, he gave us his explanation of time being our 4th dimension. This was helpful in understanding how relativistic effects impact time and distance equally.

According to quantum theories, space and time dimensions have minimum measures. Space and time are grid-like. Their minimum measures are called "Plancks", named for Max Planck who made the first discovery of quantization in 1900. It was then that he found energy to have a minimum quantity. Today we call that minimum quantity of energy "Planck's Constant." For our purposes, we'll be discussing Plancks of distance and time. The accepted size of a Planck of distance is 1.616255×10^{-35} meter and a Planck of time is 5.391247×10^{-44} second, according to the National Institute of Standards and Technology.

These measurements are extremely small. Our smallest atoms (hydrogen) are 6.56×10^{24} (that's 6.56 trillion trillion) Plancks in diameter. Even the nucleus of hydrogen, which is just 1 proton, is 6.187×10^{20} Plancks wide.

We have also found that our universe is expanding, so we'll consider how that could affect spacetime, too.

And, we should determine if space is a reliable environment. We've learned that the world is "knowable" on Earth. Our technology is evidence of that. We've made precise predictions of stars, planets, galaxies, and other bodies deep in space and our laws of physics appear to be extremely consistent. They seem to be the same in all places and at all times, since we know we're looking into the past when we look at space. But let's go a little further in establishing the structure of spacetime before we decide how we can hang our reality on it.

Could Space Flow?

Could we have this consistency of our physical laws if space was fluid?

A star emits light equally in all directions (assuming internal activity is the same throughout); same wavelengths, same intensity or quantity of photons, all traveling at the same speed in all directions.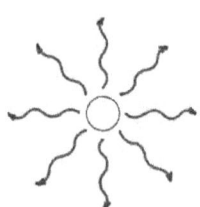

This seems to be how we experience our sun as well as all other stars which are generally fixed in relation to us.

The differences we see in the light we receive from stars are their dimness and wavelength. This is due to their distance and movement toward or away from us, because it is altered from the light they actually emit.

Suppose a star is moving away from us, and as it travels through space, the space itself is moving out of its way. The space, the Plancks themselves, flow like a liquid.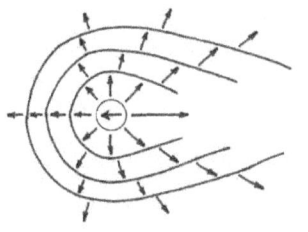

Light travels straight through space unless it's affected by gravity to redirect it (according to the general theory of relativity), or a particle to absorb, reflect, or refract it such as what we see in our atmosphere or with clouds of particles in space.

So, if space was flowing and light was going straight in relation to it, then light would be flowing, too. Light coming out of the front of a star would be dragged around to its side. The faster a star is moving, the more the light would be carried around to the side of the star and redirected. Light from the front of the star would be redirected more to the side the faster the star moves. Any observer in front of the star would see it dim according to its speed.

Likewise, an observer behind the star would see it as brighter, the faster it was traveling away. The flowing space would redirect the photons to behind it and oriented haphazardly (and at varying speeds). A moving star viewed from the side would appear as a point or sphere with a trail of turbulence.

Movement of Observer

Additionally, if the observer is moving, then space would flow around them as well, distorting their view. Looking ahead in the direction of movement, the observer's view may be most clear and free of distortion. Looking to the side, the world would appear to bend or stretch around to the observer's sides.

The faster the observer moved the more he would be able to look to his side and see what's in front of him. All of the space in the forward direction would present a fisheye view.

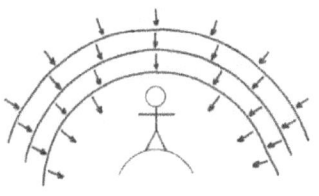

And all that's to his sides would compress toward the back. Finally, to the rear would be a mass of turbulence, a changing and unintelligible view, since all things which flow have turbulence, and turbulence is a phenomenon we haven't been able to predict the behavior of. All things which flow develop unpredictability.

We don't notice those effects as we observe stars, including those in fast orbits which have them coming toward us fast, and going away from us fast, alternately.

We see them change between redshift and blueshift, but don't notice a significant brightness difference or visible trail of turbulent space. And, we don't see space any differently in various directions we may be moving toward or away from. Our view of the sky isn't

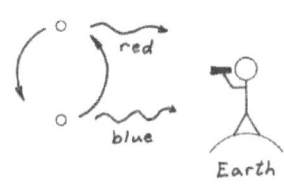

clear or blurry at different times of the day as our planet rotates. Astronomers don't notice a clearer or blurrier view in different directions. (This really only means we're not traveling very fast.)

Also, if space was getting out of the way of a moving body,

then the body would not be occupying space and what would be the location of that body? It seems to me that space does not flow, even if Earth is still.

Universe Is Expanding

Vesto Melvin Slipher, working at the Lowell Observatory in Arizona, had begun making spectrographs of cosmic systems and distant galaxies. Instead of simply looking at the light from them as it appeared, he separated it into its colors to see what wavelengths we were receiving from them.

By analyzing the wavelengths that were missing from the chart, he was able to determine the elements in the atmosphere of the star. Each element absorbs specific wavelengths.

By 1912, he had done this for four systems. While identifying the elements that the light passes through (the atmosphere of the stars), he noticed that those stripes on his spectrograph chart (which looks like a barcode) did not appear at their expected frequencies. For three of the systems those identifiers appeared in the proper arrangement but shifted toward lower energy, or the red end of the spectrum. For one system, the Andromeda galaxy, it appeared shifted toward blue, higher energy.

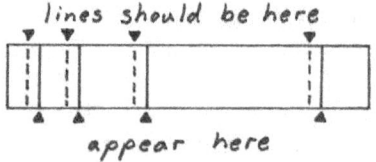

These are the spectral lines of hydrogen, the most common element making up star atmospheres.

This shift is due to something known as the Doppler effect. In 1842, Johann Christian Doppler discovered that wavelength shortens between its source and the observer when they're moving toward each other and lengthens when they're moving away from each other. The number of waves stay the same, but each one has to be longer or shorter (occur more or less frequently) to span the distance because the speed of light is constant.

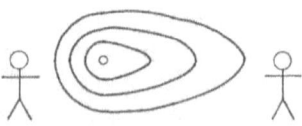

This shows itself in changing frequencies of sound as a car passes us as well as in frequencies of light (colors) as A.H.L. Fizeau discovered.

Measuring color shifts was already being done for stars in our galaxy, but not for other galaxies because, at that time, galaxies were believed to not be moving.

Slipher's discovery meant that our galaxy and the Andromeda galaxy appeared to be moving toward each other, while the others appeared to be moving away from us, or us from them.

By 1914, he had analyzed 12 more galaxies and found 11 of those to be red-shifted. And, by 1917, Slipher had found a total of 21 galaxies red-shifted, seeming to be moving away from us, and only 4 blue-shifted (or moving toward us). The average speed of galaxies seemed to be 2 million km/hr, or 0.000185 of the speed of light.

Since the wavelength/frequency shift is directly related to the speed compared to light speed, the frequencies were 0.000185 lower than normal. Wavelength (λ) consequently, was nearly 0.000185 longer than normal. It may be calculated as $v/(c-v)$. The reason for this is that wavelength grows to infinity as speed approaches light speed, while frequency (f) approaches zero.

The Doppler formula for blue-shift is $\Delta \lambda / \lambda = v / c$ ("Δ" means change). For red-shift it is $\Delta f / f = v / c$.

By 1929, Edwin Hubble had assembled findings from Slipher and others. He concluded that the universe appears to be expanding. Galaxies twice as far away were moving twice as fast, and those 3 times as far away were moving 3 times as fast. Most every system seemed to be moving away from each other. The expansion was evenly distributed.

Today the concept that the universe is expanding is generally accepted and is attributed to the Big Bang, a theory that the universe originated as an explosion, which continues even today. The rate of expansion is currently estimated at 73.04 km/sec for every 3.26 million light years of distance (according to Adam Riess, based on data of 79 stars collected between 1980 and 2021). That means a galaxy 3.26 million light years away from us is moving away at about 73.04 km/sec. A galaxy 6.52 million light years away is moving away at 146.08 km/sec. And, so on. This is referred to as Hubble's constant and is symbolized by "H_o."

(3.26 light years is a common astronomical unit of measure. It's referred to as a parallax second, or "parsec" and is a parallax of one second of arc. So, Hubble's constant is 73.04 km/sec per megaparsec.)

Hubble's constant is recalculated as new measurements are made. Some astronomers base it on measurements of distance of stars, determined by brightness, combined with the color shift from the Doppler effect and reach conclusions such as Adam Riess did. Others analyze the cosmic microwave background radiation, which is believed to have been emitted by the Big Bang an estimated 10 billion years ago or when the universe was only 379,000 years old. They find that they are able to measure the distance which sound waves have traveled through it, and calculate a rate of expansion from that. Using this method, a recent calculation shows the universe to be expanding at 67.49 km/sec per 3.26 million light years.

A simple way to visualize this expansion is to picture about 150 billion galaxies all moving away from each other, with those furthest out having to move the fastest and those near the center of the expansion not having to move as fast. And perhaps the center of the expansion is not moving at all. Whatever the actual movement of the galaxies is depends on how expansion occurs.

So, as we look up at the sky, stars within our galaxy remain near us because of gravitational forces between them and us, and the massive black hole at our galaxy's center. Those stars appear white. They move around, but not generally toward or away from us. When we see another galaxy in the distance, which is not strongly attached to us by gravity (such as those which are not a part of our cluster of galaxies), it is likely to appear red-shifted due to it moving away from us.

This is not a color shift noticeable to our eyes. It is very slight and must be measured precisely for nearby galaxies.

The light which comes to us from those distant galaxies is not just a different wavelength. It's also a different energy. Wavelength and energy are inversely proportional. Red light has a long wavelength and low energy. Blue light has a short wavelength and high energy.

If we were in one of those galaxies far out in the universe, as our planet raced in one direction and the stars of our galaxy raced with us in the same direction, the stars' movement would create a blue-shift to their light coming our way. But our moving the same direction as them, and away from them at the same speed they're coming toward us, undoes the blue-shift by creating an equal red-shift. Our stars ahead of us speed away creating a red-shift, while we speed behind them at the same rate creating an equal blue-shift. And all of our galaxy's stars which we're traveling with appear normal, white.

The galaxies behind us, which we would be moving away from, would appear red-shifted. And the galaxies ahead of us, which we're moving toward, would be moving away from us faster than we're moving toward them. So they would appear red also, same as those behind us.

If we were far from the center of expansion, this color shift wouldn't tell us anything. If we were at the absolute far reaches of the universe, where the galaxies are moving fastest, our stars would appear normal to us because we're traveling with them.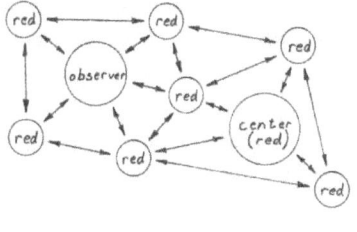

Looking toward the center of the universe, all galaxies would appear red-shifted as we move away from them, assuming the universe is expanding equally. In the other direction we would see no galaxies beyond our own. We would only see our own galaxy's stars. We would be at the outer edge, if the universe has an edge.

There is an obstacle to the expansion of the universe and that is the cosmic speed limit, the speed of light. As Einstein said, nothing can move faster through space than light. And the speed of light is fixed at 299,792.458 km/sec (670,616,629 mph). With that limit in mind, and the expansion rate of 73.04 km/sec per 3.26 million light years of distance, we can determine that any object that is 13.4 billion light years or more from the center of the universe is moving at nearly the speed of light.

$$\frac{299,792 \text{ km/sec} \times 3.26 \text{ million light years}}{73.04 \text{ km/sec}} = 13.4 \text{ billion light years}$$

Beyond 13.4 Billion Light Years

When a body is moving away from the center of expansion at or near the speed of light, it can go no faster. If we lived there, we would not see distant galaxies twice as far away moving twice as fast when we looked toward the galaxy's edge. As we looked away from the center of expansion, we would not see any of them moving away from us. We would be moving at the maximum speed and so would they, at the same speed. All galaxies beyond 13.4 billion light years away would travel together, and they would show no color shift from the Doppler effect, in the outward direction.

Because we would be moving outward from the center, we would also be growing further from each other in the directions perpendicular to the center, sideways.

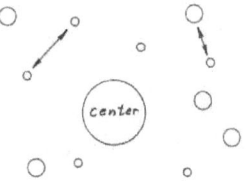

This is the speed relative to the 2 galaxies, as a result of absolute movement. If all galaxies are moving directly away from the center of expansion, then they will show a red-shift to each other. That red-shift will be a portion of the red-shifts of each of the stars. And the shift will be determined by the angle they are moving away from center.

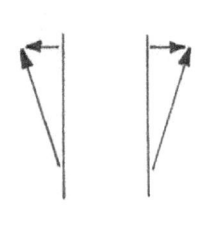

This is the rate they appear to be moving away from each other as if they had no outward movement.

Our outward speed would remain near light speed, while our distance from other galaxies perpendicularly would grow. This distance could grow faster than the speed of light since it's not the speed of the galaxies through space. If our galaxy were in this situation, we would see galaxies all around us red-shifted except those directly ahead of us in our near-light-speed travel. Those would be white, and that would tell us we're not at the center of expansion.

While it may seem that no place is special in the universe, and all locations may be equal, there does seem to be a center of expansion, or at least a central area of 13.4 billion years radius in which expansion is occurring at less than the speed of light.

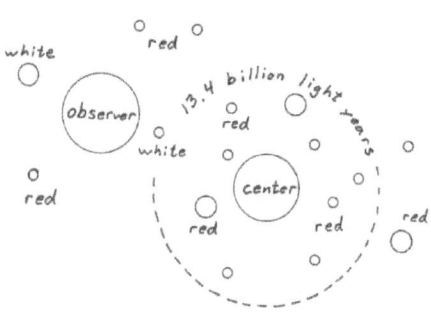

From well within this area, the universe would appear the same all around. It would appear from that special location that we weren't in a special location. It would appear that all was expanding evenly, and perhaps that space itself is growing. (This is the view we do have. The universe does look the same to us in all directions.)

Simply considering the expansion of the universe in this way, we see how the cosmic speed limit prevents objects from accelerating indefinitely. That is unless space is curved into a ball as Riemann suggested.

A Curved Universe?

Einstein used warping of space and time to explain the dilation effects from relativity as well as how gravity works, that gravity is not a force but a warping of spacetime. So he used the three-sphere, or spherical space, model developed by Reimann to explain that our space is curved into a ball and so it has no center, and that is why it appears the same in all directions. The cause of this curvature is primarily gravity.

In 1917, in "Cosmological Considerations on the General Theory of Relativity", Einstein wrote: "According to the general theory of relativity the metrical character (curvature) of the four-dimensional space-time continuum is defined at every point by the matter at that point and the state of that matter."

The state of the matter may involve velocity, but we know he believed the universe is static, so the majority of the curvature would be from gravity.

It's this curvature which allows for worm holes to exist (two black holes connected by a tunnel) and also for a traveler to be able to return to his starting place by going far enough in any one direction.

Curvature is said to eliminate the need for a cosmological constant. Instead, the universe expands similar to the way a balloon inflates so that galaxies move away from each other equally. The universe gets bigger, but there's no center and no outer edge. While the analogy is that it's like a balloon, we can't actually visualize it in our 3-spatial dimensions. We must imagine all 3 of our spatial dimensions as if they were the 2-dimensional surface of the balloon.

This model requires that space itself is growing. Because if space is expanding, then the galaxies don't have to move away from each other, and no force (cosmological constant or dark energy) is needed to push them apart. Instead, something to further inflate the balloon is needed.

Because space is quantized, there are 2 ways for it to grow without moving the galaxies. One is to stretch the grid of space, as the balloon analogy implies. The other is to continually insert Plancks of space at a specified rate.

Stretching Plancks

By stretching Plancks, the grid spaces themselves, which is our unit of measure, would get bigger. 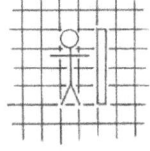 Our method of measuring distance in space, which is light, would be unable to detect a stretching of grid spaces. In our grid, light moves 1 Planck of distance for every 1 Planck of time. (This is the relationship between the Planck units of space and time.) If the Plancks of distance were stretched, light would still travel 1 Planck of distance for 1 Planck of time. Our  measurement device, light (or even a tangible device like a ruler), would stretch the same as space and any change would be undetectable. The only way we could notice a change in distance is for more grid spaces to be added.

Adding Plancks

The way to insert new grid spaces into our world and create the equal expansion everywhere, which we appear to have, seems to be to add the grid spaces equally everywhere. This would effectively push everything apart without requiring the movement of objects. But every object would get pushed away from, or grow more distant from every other object. This includes every subatomic particle.

In order for the parts of all of our stars, planets, molecules, and atoms, to not disperse as a thin gas throughout space, natural forces would pull them together.

The expansion of space would create movement of almost every particle just to maintain the universe of systems which we have now. Nuclear forces would pull quarks inward to hold together atomic nuclei. Electromagnetic forces would pull electrons constantly toward their nucleus to keep atoms and molecules together. And gravity would create constant motion to hold planets, stars, galaxies, and galaxy clusters together.

The speed of all these things through space, just to stay in place, would seem to be based on Hubble's constant. This is a very very small speed when describing a person constantly moving downward to remain on the Earth and not float away, which becomes much greater with larger systems.

However, it would actually be a far greater speed. So far, we've been relying on inertia to keep objects at rest in the space where they're at. Inertia is a resistance to acceleration, but it doesn't stop acceleration. Forces such as gravity pulling objects toward each other are accelerating forces.

In the first moment of this arrangement, adding space works perfectly. The universe expands. But since the bodies are free to move through space with virtually no resistance, they would not only begin to fall toward each other, they would continually accelerate toward each other no matter how much inertia they have or how slight their attractive force is.

Their rate of acceleration would be based on that gravitational attraction (or other natural force), which is expressed as speed per second. And the seconds for which they have been accelerating toward each other could be as much as the age of the universe. If the universe is 13.8 billion years old, as is believed, then galaxies may have been gaining speed for 4.35×10^{17} seconds. Space would have to grow at an accelerating rate to keep up with the falling. The gravitational attraction between two galaxies could be multiplied by 4.35×10^{17} to tell us their current rate of fall toward each other. Space would have to expand at that rate for the universe to remain static. Then Hubble's constant would be added to that in order to create the expansion our universe has. Moving through space at those kinds of speeds would result in a dramatic Doppler effect, not to mention other effects. Inertial resistance along with expanding space cannot easily take the place of a force.

To avoid all that, so that we may proceed, we need to have a repulsive force to counter the falling. It must be constant and equal to the attractive force between major systems. With this force in place, let's proceed.

We saw how objects at the far reaches of the universe approach light speed based on Hubble's constant. The universe itself is the largest object of all, but it doesn't seem that it's being held together. It's expanding.

Our galaxy clusters and galaxies, however, are impressively large, too. The supercluster of galaxies we're a part of is about 200 million light years in diameter. It is the biggest system which holds itself together that we're a part of. The gravitational centers of these largest systems would be the only things not moving in space. All other parts of the supercluster would have to move toward it, then, to stay a part of the system. The outermost galaxies of our supercluster would have to constantly move inward at 4,800,000 mph.

Our local cluster of galaxies is about 10 million light years in diameter. And our own galaxy, the Milky Way, is 100,000 light years in diameter, with Earth being about 27,000 light years from its center. Remaining a part of our own galaxy, local cluster, or supercluster could require significant speed by our planet moving through space if we're not at its center.

In addition, if we are to be in motion relative to the center of the supercluster, then we are also in motion within our local cluster, moving toward Andromeda at over 137,000 mph, traveling through our galaxy at 450,000 mph and orbiting our sun at 66,900 mph. The only way all these movements would not occur would be for us to be at the center of the supercluster and for all bodies to be moving around us, to create the movement we see, and inward toward us to hold together.

Having eliminated falling speeds in a way which keeps galaxy clusters at rest in space, if we are to be traveling only at these speeds, we still have relativistic effects. We'll determine, later, whether we could be experiencing the effects of relativity from the orbit we appear to have around our sun, throughout our galaxy, and movement toward Andromeda within our cluster of galaxies.

Additionally, with expanding space, there's a problem of maintaining the speed limit of light. Light speed would vary with distance. There would be no fixed speed of light. It would be different for every distance of travel. The speed of light from one end of this room to the other would be greater than the speed of light across this table. And, even greater would be the speed of light from the sun to the Earth. This would be the result of distance continually growing.

The reason is because the universe is expanding per second per distance. So, it would compound. Simply growing per second or per distance would create a greater speed. Both together create an exponentially increasing speed.

Light coming from the sun to the Earth travels 150,000,000 km at 300,000 km/sec, so it takes 500 seconds to get here. With that amount of space growing at Hubble's constant for 500 seconds, that light will have traveled (or be credited with traveling) an extra 1.755 x 10^{-10} of light speed, or 3.507 x 10^{-10} km/sec.

Light coming from the Sagittarius galaxy takes 78,000 light years to get here. In that time, the distance will have grown by 0.449 light years. And the speed of that light would measure an extra 5.75 x 10^{-6} of light speed, or 1.7 km/sec faster.

This could be compensated for by adding Plancks of time equally with those Plancks of distance. That way whatever the distance traveled by a beam of light, it will have always gone at light speed, 1 Planck of distance for 1 Planck of time.

But in doing this, history would be created which was not experienced by anything. Anything moving, other than just to maintain its place in a gravitational system would have a history of distance and time which is greater than the actual distance and time it traveled. This would work against the dilation effects of relativity, though it would be less. It would cause the decreased experience of time and distance lost due to dilation effects of travel to be increased by space growth. The amount of this increase would be Hubble's constant, and it would cause predictions of relativity to be off by that amount.

That amount is 73.04 km/sec per 3.26 million light years, which is 2.339 x 10^{-18} m/s compounded every meter of distance.

This leaves us with 4 issues:

1. Constant movement to maintain our place in the super-cluster of galaxies, as well as our other movements throughout our solar system, galaxy, and supercluster. These would all create relativistic effects which we

should be able to detect as distance and time dilation, if movement is absolute. (If movement is only relative to each observer, then this is not a concern.)

2. All relativity predictions would be countered slightly by Hubble's constant.

3. What would be the cause of new Plancks of space and time being added, this continuous act of creation of a history which was never experienced?

4. We do need a repulsive force to avoid all bodies falling inward toward each other, accelerating to very high speeds. And, it must match the amount needed to keep the centers of galaxy clusters at rest in space plus our rate of expansion.

It also relies on a source of curvature, if the universe is spherical, something to determine the amount of curve, and therefore the size of spacetime. Einstein believed relativity created the curvature, gravity, and the relativistic effects from travel.

In absolute relativity, I suggest that only experience gets reduced by travel, not actual distance and time. And, I see that gravity appears as a force, not a warping of space and time. This leaves us with no reason to believe that space and time are other than what they appear to be, flat and fixed.

Is There Only One Fixed Grid?

Suppose I get some Lego blocks and use one of the set's broad platforms to mount my pieces on. I build a house on one part of it. Then, nearby, I build a garage separate from the house. Because of that grid (the platform) I built both of them on, I know I can go back later and add a covered walkway between the two which will fit. Lego blocks come in standard measures of length and so does their platform. They are each quantized and they use the same size units. Our world does the same.

But, what if I hadn't used that broad Lego platform (grid) to build both my house and garage on? What if I just built my house on the table, on empty unquantized space, then built my garage nearby?

If I wanted to use my quantized Lego bricks to build a covered walkway connecting them, I certainly would have to move one of my structures in order to make the pieces fit together.

The same would apply to my real house. Even though we know our world is actually in units of measure which are super tiny (Plancks), let's imagine the unit of measure is a foot. So, I build my house 10 feet from the side of my property and 30 feet back from the sidewalk. Then I build a garage next to it, 6 feet away. I make it an even 6 feet so later I can build a covered walkway to join them, since I can only cut building materials in even feet. If I had built the garage 6 1/2 feet away I wouldn't be able to connect them with a covered walkway later.

You live very far from me, so you don't have to worry about my grid, or coordinating your measurements with mine. Until you decide to visit me. Then, since you must travel in even feet only, you travel a million feet to get to my house, or near my house. You find that my house is still about a half of a foot away. We both have the same one foot minimum distance of measure, but without a common grid, we don't align. You find you're unable to knock on my door because you can't even touch my house. You try yelling, but the sound waves are in minimum measures of feet, too, and can't get to me. You write a note before leaving, but have no place to put it, since I can't pick it up if it's a half of a foot away from me. And, I can't see it because, like sound waves, light waves are also in minimum measures of whole feet.

So, it seems that if there were multiple grids they would have to coordinate, meaning there really is only one grid. Just because we can't see the grid, as we can with Lego blocks, doesn't mean it's not there. And, I don't believe that just because Plancks are small they should be imprecise.

Space and Time Conclusion

The original problem I set out to solve was the small discrepancies in our realities which emerge from relativistic effects. Since the differences are very small, the problem seems to be very small.

Considering that we each may have our own grid in order to accommodate the relativistic discrepancies seems to reduce the problem to less than Planck size. They are the smallest of all measurements, and therefore we seem to reduce our problem to be the smallest of all problems, the fact that they do not align.

But I don't think making a problem smaller makes it go away, even if it's made as small as possible. I believe the only logical solution to the misaligned multiple grids issue is that there is only one grid.

I did leave the expanding space possibility somewhat unresolved, for now. Because of the explanation required, I'll get back to it later. But, I'll tell you ahead of time that I don't think expanding space provides a viable explanation for how the universe expands. You may decide for yourself when we return to the question.

For now, I'll proceed with why I believe movement is absolute.

Chapter 6 - Time Dilation

I believe we can determine absolute movement using time dilation. Through relativity we've learned that the laws of movement as well as all other laws of physics are relative between the observer and the observed. This has been well established by experiment.

My suggestion is that the way they are related to each other is through their relationship with spacetime. And I believe we prove it all the time.

Here's a thought experiment using time to show how this works. The setup is a little unusual, but it is to avoid bias in thought as well as potential complex influences of being on Earth.

Imagine that you and I are each floating in space and experiencing no detectable gravity or acceleration. We can't tell whether we're moving or not. Except for one thing, we keep passing each other.

I see you approach me from my left and leave to my right.

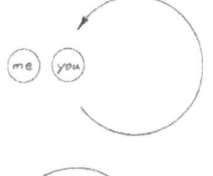

You see me approach from your left and leave to your right.

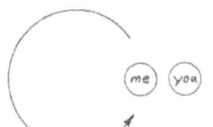

We each see the other traveling in a large circle. One of us is likely orbiting a black hole or dark matter so that its gravity and our centrifugal force (outward force from orbiting) cancel each other and we can't feel it. It's why astronauts on the space station feel like they're floating in space even though they're traveling around the Earth at 18,000 mph.

The special theory of relativity tells us that movement is relative to the observer. So, since you're moving compared to me, then to me you should experience less time than me. But, since I'm moving relative to you, then to you I'll experience less time. As we pass each other we synchronize our watches. Then the next time we pass we can compare them. When we compare, we find out if our movement is only relative to the observer or if there is something we're all relative to, or related to each other through. Is there absolute movement independent of our own experience?

If everything is only relative to each observer, then I will see your watch read 12:59 while mine reads 1:00, because you're the one losing time by moving. And you will see my watch read 12:59 while yours reads 1:00 because to you I'm the one moving so time moves slower for me.

Could we both look at the same watches and see different times?

Suppose a meteoroid demolished my spacecraft just before 1:00. If you see me at 12:59 my time, I'm still alive. However, because I don't reach you until 1:00 my time I never see you because I die before I get there. When you see me at 12:59 my time, before I die, we have a conversation. Or, you have a conversation with me, but I don't have one with you because I died before I could see you.

What if while you meet me alive, you warn me to get out of the way of the meteoroid? So, to you I live. But to me I die because I wouldn't meet you until 1:00 my time, after the meteoroid kills me.

Am I alive in your world and dead in mine?

Different Grids

A solution might be to say that we agree on all the events, just not the times they occur. If time is a dimension like space is, then maybe we can each have our own. If we're living at different times, then what happens at 1:00 my time could happen at 12:59 your time according to me. And, while it happens to you at 12:59 your time according to me, it would happen to me at 12:59 my time according to you, while it's 1:00 your time.

If we were playing basketball, I would be telling you the game's over, but because you see the clock as being slower you

48

would then argue for more time. Actually, I would be telling you the game's over while you hear me arguing that it's not. You could be telling me it's over while I hear you arguing that it's not. It has us in different time dimensions which can't coordinate.

While it may seem that we could just translate our times, that's not so easy. Simply determining that 12:59 your time is 1:00 my time could work, except that 1:00 your time is 12:59 my time to you, which is also 12:58 your time to me, which is 12:57 my time to you. The reason we can't even agree on our disagreement is because we both experience dilation in relation to the other person, not just one of us.

The reason I said we prove all the time that there is absolute movement is because we never run into these situations. And, the slightest discrepancy in our real world would show itself in technology and biology. The same mismatch of realities we see in this space experiment of a minute difference would also result from very small discrepancies in the superfast transactions of computers and the chemical reactions of our bodies.

Whether a transaction occurs before or after another one could have dramatic consequences while computing information or executing life processes. Chemical processes can occur or not occur in as little as a trillionth of a second. Your experience of time could become less than your spouse's by that much with a trip to the grocery store. (35 mph for 12 minutes reduces your time experience by a trillionth of a second.) Could you and your spouse disagree about becoming pregnant? What about whether you've spent the last 20 years raising children or not?

We can be sure that these small differences would occur because the relativistic effects are always there. There is always a reduction of time and space due to movement, no matter how slow.

When you walk at 2 mph you're losing 4.5×10^{-18} seconds (0.0000000000000000045 seconds) for each second you're moving.

If a snail could go 12 inches in a minute, it would lose 1.4×10^{-22} seconds for every second it's moving, or 8.4×10^{-21} seconds total for the trip.

While we say relativity doesn't affect us at our normal speeds because as humans we can't notice the tiny effects, it is necessary that our molecules don't have their own slightly different realities. They, like computers, need to agree on one reality for their processes.

Going back to our space experiment, let's consider our one reality and see what it tells us.

We look at our watches and we agree. Mine says 12:59 and yours says 1:00. I experienced less time in our one reality. So, while according to me <u>you</u> were moving, and according to you <u>I</u> was moving. The truth is, <u>I was actually moving</u>. And, the only thing to be relative to then, is spacetime itself.

Understanding how this works can help us expand our knowledge of the world at various scales, especially those involving significant speeds which give us time differences we can easily measure.

Since we know that all movement is relative through its relationship with space, and we know we can use the effect of time dilation to detect it, we can now determine our own movement. How the Earth is truly moving. We have the data for it already.

The Absolute Movement of Earth

Remember that in 1971 Joseph Hafele and Richard Keating flew cesium clocks around the world to test the effects of relativity on time. The loss of experience of time is caused by both movement and gravity. Both were involved in the experiment's calculations and the results proved predictions to be correct. So, lets deconstruct this experiment and find out what we can about Earth's movement in relation to space. (We can do this because Joseph Hafele and Richard Keating both looked at their clocks and agreed on the times they saw.)

This experiment involves time changes from both gravity and speed. The speeds involved are both the speed of the plane, and the speed of the Earth.

50

Gravity slows time more at the surface of the Earth than it does at the altitudes our planes fly at. This is because gravity is stronger when closer to the center of mass. Since gravity is weaker higher up, a clock on a flying aircraft will run faster. We consider this as a gain based on how high the plane is.

Both the aircraft clock and the Earth clock actually lose time due to gravity, but the aircraft clock loses less. The amount of the relative gain is 10^{-16} seconds for every meter above the Earth. At one meter, that clock will advance by 1×10^{-16} second for every second which passes, compared to a clock on the ground. At 2 meters altitude, the clock will gain 2×10^{-16} second for every second which passes. And the faster the movement is, the slower time is. So, the faster a plane flies, the slower the clock on it runs.

While we are taking a shortcut in the way we're calculating gravitational dilation, we can't do that with speed. For the effect of gravity we only need to consider altitude. For speed, we've got multiple possible speeds and directions involved. The Earth may or may not have various motions. That's what we're aiming to find out.

Mechanically, we know that the Earth rotates each day. And since it's 24,906 miles around at the equator we can say it's going 1038 mph at that latitude. This is a significant speed and will definitely affect measurements taken in either direction by changing our absolute speed through space.

Earth orbits the sun at 66,900 mph. If this is part of Earth's absolute movement through space it will make a big difference in the results. In addition, we orbit our galaxy at 450,000 mph. That would show up even more dramatically in our results.

Dilation effects become more dramatic as speed increases. Even small differences at higher speed ranges experience amplified dilation.

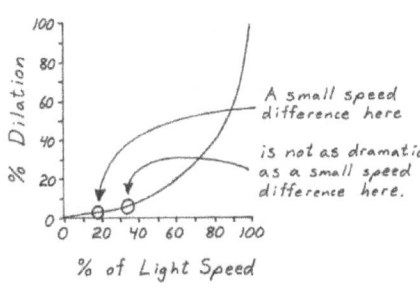

% of Light Speed

Hafele and Keating flew their clocks around the world east and also west using multiple connecting commercial flights, and not straight around the equator. So, there are many factors to consider which would affect the time the clocks displayed when they finally got back to Washington, D.C. where they were compared to a stationary clock at the U.S. Naval Observatory.

Their results, again, were:

- The clock flown east lost 59 nanoseconds.
- The clock flown west gained 273 nanoseconds.

(A nanosecond is a billionth of a second.)

While the details of their data are complicated by multiple takeoffs and landings and various distances traveled at different altitudes and speeds, the results are still able to tell us about the true movement of Earth in space.

I found the results to be consistent with flying east at 534 mph at an altitude of 35,000 feet for a distance of 20,050 miles. (The circumference of the Earth at Washington, D.C.'s latitude is 19,383 miles.) Going west, I found their result consistent with flying 379 mph at 31,900 feet for 19,383 miles. To get my results to match Hafele and Keating, I had to account for the Earth's daily rotation, but not its orbit around the sun.

Had the Earth not been revolving in relation to space, the results would have been:

- east - 101 nanosecond gain
- west - 150 nanosecond gain

Had the Earth been both revolving and orbiting the sun, the results would have been:

- east - 10,823 nanosecond gain
- west - 3,861 nanosecond gain

Both of these scenarios predict dramatically different results from what the experiment shows. It's plain to see that the Earth does revolve each day but has no significant movement otherwise.

The next few pages go over my method of calculating. You may skip them if you'd like, since I've already provided the answers. But you may also explore them to verify my work or pursue your own analysis of data. Skipping over the math should not affect your understanding the message of this book.

Dilation Formula

The formula I used for dilation (the loss of experience) is:

$$1 - \sqrt{1 - v^2/c^2}$$

The function Lorentz formulated, and Einstein used was:

$$\sqrt{1 - v^2/c^2}$$

v = travel speed
c = the speed of light

I've subtracted the result of the Lorentz formula from "1" to give me the portion of change of experience rather than the amount of the whole experience. This way it leads me to time differences instead of total time experienced during the experiment. (Einstein did this too, in order to determine the amount of change in time.)

While many readers may be most familiar with speed and altitude expressed in miles or feet, I have used all metric measure for these calculations.

The Earth clock used in the experiment at the U.S. Naval Observatory in Washington, D.C. is located at about 38.9° latitude. At this latitude, the circumference of the Earth is 31,187 km (19,383 miles). This means that as the Earth rotates, this clock moves 1299 km/hr, which is 0.000001204 of the speed of light.

At this speed, the Earth clock experiences 7.249 x 10^{-13} dilation (loss) of time.

Eastbound Flight Calculations

When flying east for their experiment, Hafele and Keating predicted a loss of 184 nanoseconds due to speed differences, and an effective gain of 144 nanoseconds due to altitude. This was a total loss of 40 nanoseconds of time by the flight clock compared to the ground clock.

For the eastbound flight experiment, I found the following data to give me a match for their predictions.

- altitude of 35,000 feet (10,668 meters)
- flight speed of 534 mph (859 km/hr) going the same direction as Earth's rotation (1344 km/hr along the flight path) for a total absolute speed of 2203 km/hr
- flight distance of 20,050 miles (32,260 km)
- flight duration of 37.55 hours

Time Loss Due to Speed

Based on total speed for the duration of the flight, the flight clock should have lost 281.6 nanoseconds, while the Earth clock lost 98.0 nanoseconds. This would result in the flight clock showing 184 nanoseconds less time.

Time Loss Due to Gravity

Based on the altitude and flight duration, the flight clock should show 144 seconds greater time than the Earth clock.

Net Result

Combined, the flight clock should show 40 nanoseconds less time than the Earth clock after circumnavigating the globe going east.

The actual result of Hafele and Keating's eastbound trip was a loss of 59 nanoseconds. This difference could possibly be accounted for by inconsistencies in speed or distance, which are difficult to measure in flight compared to altitude or time.

Westbound Flight Calculations

On their westbound flight, Hafele and Keating predicted a gain of 96 nanoseconds due to speed differences, and an effective gain of 179 nanoseconds due to altitude for a total gain of 275 nanoseconds of time by the flight clock compared to the ground clock.

For the experiment flying west, I found the following data to give me a match to their predictions.

- altitude of 31,500 feet (9725 meters)
- flight speed of 379 mph (610 km/hr) going the opposite direction as Earth's rotation (1299 km/hr along the flight path) for a total absolute speed of 689 km/hr (going EAST, not west)
- flight distance of 19,383 miles (31,187 km)
- flight duration of 51.13 hours

Time Loss Due to Speed

Based on total speed for the duration of the flight, the flight clock should have lost 37.5 nanoseconds, while the Earth clock lost 133.4 nanoseconds. This would result in the flight clock showing 96 nanoseconds greater time than the Earth clock.

Time Loss Due to Gravity

Based on the altitude and flight duration, the flight clock should show 179 nanoseconds greater time than the Earth clock.

Net Result

Combined, the flight clock should show 275 nanoseconds more time than the Earth clock after circumnavigating the globe going west.

The actual result of Hafele and Keating's westbound trip was a gain of 273 nanoseconds.

Orbiting The Sun

Earth's orbital speed is 29.8 km/sec. This is the speed the Earth needs to go in order to get around the sun once per year, and maintain its 149,600,000 km average distance from it.

If we were to look down on the Earth from the north pole, or well above the north pole in space, we would see the Earth rotating counterclockwise and also orbiting around the sun counterclockwise.

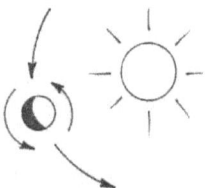

This means that if we want to consider the speed on Earth in the middle of the day, we would subtract the rotation speed from the orbit speed. In the middle of the night, we would add them.

For this experiment, I assumed daytime flights, which is the smaller, less dramatic number, but also the most likely.

I took the travel speeds I used previously for the plane and the Earth clocks and added orbital speed to them.

East

Flying east, we're going with the Earth's rotation direction, but against its orbit direction during the day. So, we use the plane speed combined with rotational speed against orbit speed, and get 29.20 km/sec. This gives us a loss of time of 640,753 nanoseconds for the flight clock.

Meanwhile, the Earth clock lost 651,753 nanoseconds due to its speed. So, the flight clock will appear 10,679 nanoseconds ahead of the Earth clock due to speed.

Gravitational effects on time remain the same as in previous calculations, giving us a total net gain of 10,823 nanoseconds by the flight clock.

West

Flying west, we're going against the Earth's rotation, but with its orbit during the day. So I used the plane's speed, less the rotational speed, combined with the orbit speed to get 29.61 km/sec. This gives us a loss of time of 881,502 nanoseconds for the flight clock.

Meanwhile, the Earth clock lost 885,183 nanoseconds due to its speed. So, the flight clock will appear 3,681 nanoseconds ahead of the Earth clock due to speed.

Gravitational effects on time remain the same as in previous calculations, giving us a total net gain of 3,861 nanoseconds by the flight clock.

Could Space Be Expanding?

As we explored the model of an expanding universe in which space itself is doing the expanding, we found that for the model to work, all parts of systems would have to continually move through space toward the center of the system just to avoid drifting away.

The result of that would be that space would seem to only grow in-between the gravitational systems.

The Earth could not be fixed in space in that case. All of its movements through the solar system, galaxy, cluster, and supercluster would be occurring in relation to space in addition to its move toward the center of the supercluster.

These speeds would be:

- orbiting the sun - 107,000 km/hr
- orbiting the galaxy - 724,000 km/hr
- moving toward Andromeda in our cluster-220,000 km/hr
- moving through our supercluster - Unknown
- moving toward the center of our supercluster - Unknown

I think that, based on the relativistic effects of time dilation which Hafele and Keating experienced during their flights we're certainly not traveling through space at those types of speeds. So, the universe cannot be expanding by Plancks of space being added equally everywhere.

Dilation Grows Exponentially

I've shown this illustration before, but I want to point out its significance. The way that relativistic effects increase, makes them a tool for measuring true movement.

When our absolute speeds for the east and west flights were 2203 km/hr and 689 km/hr. Their dilation from speed was 2.083×10^{-12} and 2.038×10^{-13}, with a difference of 1.879×10^{-12}.

When our absolute speeds for the east and west flights were 29.20 km/sec and 29.61 km/sec, their dilation from speed was 4.740×10^{-9} and 4.789×10^{-9}, with a difference of 4.900×10^{-11}.

The exponential growth of dilation causes their experience difference to grow even though the speed difference remains the same. A dilation of 4.900×10^{-11} is 26 times greater than 1.879×10^{-12}.

Looking at the illustration, we can see that as our absolute speed goes up, we move up the curve and dilation differences become more dramatic even though speed differences remain the same.

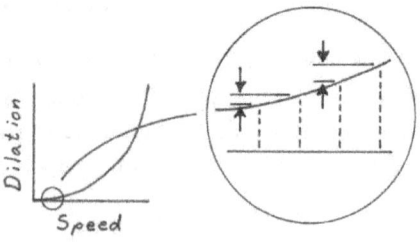

Absolute Movement of Earth

It's clear that the results of Hafele and Keating's 1971 experiment fits with the Earth's daily rotation, but not with its orbit of the sun. And definitely not with its movements around the galaxy.

Fat Earth

In support of Earth's absolute rotation is Earth's fatness around the equator. The circumference there is 40,074 kilometers while its circumference in the north-south direction is only 40,009 kilometers. This is believed to be caused by the centrifugal (outward) force from rotation being strongest at the equator. And this is evidence that Earth is actually rotating relative to space.

Our Lack of Movement Through Space

The Voyager space probes provide evidence of the lack of our movement through the galaxy.

Launched in 1977, Voyager 1 and Voyager 2 have survived well beyond their original expectations. They were only intended to explore the planets, but have continued reporting data to us all the way out of our heliosphere (the area of influence of our sun's magnetic field).

In 2012 and 2018, they passed into interstellar space and reported no sign of the sun speeding through space.

The sun emits cosmic rays (which are mostly hydrogen and helium nuclei, atoms without their electrons) at a million mph (they slow with distance) in all directions. These particles carry the sun's magnetic field with them to form what's known as the heliosphere. The outward pressure from this wind pushes back plasma from interstellar space.

It's also expected to form a shape somewhat like that of a comet, since our sun, and planets included, are believed to travel 450,000 mph through our galaxy. This would create a bending of the paths of particles coming from 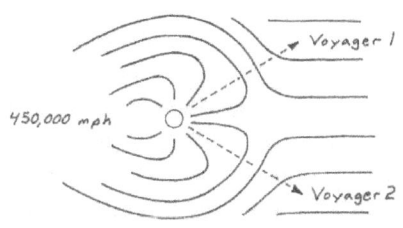 our sun as well as from other stars, as particles from each push against each other like two opposing winds. The bending of magnetic field lines should be significant as our particles collide.

59

When each of the Voyager probes passed from our heliosphere into interstellar space, they encountered the excess plasma which was expected (80 times more than that within our heliosphere). This was due to our solar wind pushing back plasma from space. But it did not encounter a change in direction of magnetic fields. The magnetic field lines only showed small variations in direction. They seemed not to be bent from high-speed collision.

They provided us with no evidence that we're moving through space in any significant way. Instead of the two winds colliding at 450,000 mph, it seems they only collided at hundreds of mph or less. (Our solar wind slows to hundreds of mph or less in relation to the sun before leaving the heliosphere, but should still travel in excess of our sun's speed in relation to interstellar space.)

Since these particles have mass, they carry with them the inertia of their source. Essentially, they behave according to Newtonian laws of motion. So, while this evidence doesn't provide direct support for absolute relativity, it does come to the same conclusion.

We've found that space itself does not seem to be expanding, and the Earth has no significant movement in space other than its daily rotation. And the absolute movement (movement in relation to space) of all other objects can be determined by their relationship with us.

I realize that Earth's apparent movement around the sun, as well as other movement, seems to conflict with this finding. We'll, also, be encountering further similar evidence later in this book. I won't be presenting a solution to this conflict, but believe one does exist. For now, I will proceed because the issue is small enough that we can still evaluate the universe.

Chapter 7 - Distance Dilation

According to relativity, distances are reduced only in the direction of movement. And, the popular understanding seems to be that distances in other directions are not.

I suggest instead, that if distance is reduced in one direction, then it is equally reduced in all directions. Distance being reduced only in the direction of movement causes distance to be reduced equally in all directions when considering absolute movement, because movement is required for all measurement.

Everything Shrinks

When an object is moving, it experiences less time and distance in all directions. If you were moving fast, everything would seem to get smaller, but you wouldn't know it because anything you could use for comparison, any measuring device, would also get smaller.

Suppose you're traveling in some direction at 292 million mph, 0.436 of the speed of light. Your experience of distance and time gets reduced by 10%.

Imagine you're in the cabin of your spacecraft, which is 10 feet long and 10 feet wide. To move yourself forward from end to end you would only have to go 9 feet. Because you're traveling 292 million mph, your own speed gets added to that as you maneuver forward in the cabin. (You can't walk because you're weightless.) It gets subtracted if you move toward the back of the cabin. However, since you only move at 2 mph, the effect is insignificant. Moving either direction, you only have 9 feet to go.

But what about sideways? If the cabin isn't moving sideways and you move inside it directly to the side, then the only speed you have sideways is 2 mph. So, it seems the cabin may still be 10 feet wide. But it's not.

Because your spacecraft is going forward, when you move directly to the side, your true movement is at an angle. In the few seconds it's taken you to maneuver yourself from side to side within the cabin, you've also gone forward quite a long way.

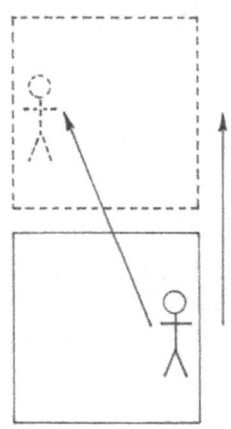

Your true movement is the hypotenuse of the sideways and forward movements.

Your diagonal, true movement is just slightly faster than your forward speed. You may calculate it as the hypotenuse with the Pythagorean theorem, $a^2 + b^2 = c^2$.

$$(292 \; million \; mph)^2 + (2 \; mph)^2 = (true \; speed)^2$$

But, as before, your maneuvering speed in the cabin is so insignificant it won't make any practical difference. Regardless of your sideways speed though, the hypotenuse will always be at least as great as your forward speed. In this case, the distance of the hypotenuse is reduced by 10%, the same as forward distance is. And as a result, so is the sideways distance.

Geometry assures us that our experience of all distances will be reduced equally according to our absolute speed in any direction.

No matter what direction you travel in this high speed (what's known as relativistic speed) reference frame, all distances are seemingly reduced by 10%. Everything seems to become smaller. This includes your ruler, and you, so you wouldn't notice. (There is a visual change the traveler will notice that occurs by a different effect. We'll get to that in Visual Distortion.) This is why we refer to the bystander observing the dilation effect on the mover. But he doesn't see it either.

Because nothing's actually smaller, only the experience is, a bystander will always see a moving object as its true size and speed.

62

If a fast-moving object were to appear smaller than it is to a bystander, what would he see to fill the extra space around it? Would everything around it, then, get bigger? Or does light bend around moving objects based on their speed, regardless of their gravity?

If a thing is experienced as being smaller, but you can't tell because you're traveling with it and you seem smaller, then you wouldn't know you're experiencing it as being smaller.

The experience is relative to absolute time and space then. The bystander is the one who knows. The one who is at rest or is slowest moving through space. He knows not because he sees you as smaller but because you tell him of your experience and he knows it was less than what he witnessed.

The smallness which is experienced is balanced by reduced time resulting in the observer in motion having a normal experience. Had distance not been reduced, only time, then he would perceive events as happening fast (in less time).

Life is different at high speeds. At least when you're not the one speeding.

To the rest of the world, fast-moving things operate slower. While their speed through space is fast, their processes are slower. This includes biological processes, chemical, and atomic processes. Fast-moving things age slower because they experience less time.

Effects of Gravity

When calculating the dilation effects of gravity in our distant galaxies, we may consider their speed also. The speed at which a body is traveling increases its kinetic energy, which contributes to its total mass/energy, and therefore its gravity. In an expanding universe, when applying absolute relativity, gravity of distant bodies can become substantial.

In 1911, Einstein wrote, "...energy must therefore possess a gravitational mass which is equal to its inertial mass." And in 1916 he wrote, "...the total mass of the system, and therefore its total gravitating action as well, will depend on the total energy of the system, and therefore on the ponderable energy together with the gravitational energy."

We know a body's gravitational force by Newton's formula of $F=Gm/R^2$, in which "F" is the gravitational force, "G" is the gravitational constant of 6.672×10^{-11} Nm²/kg², "m" is mass in kilograms, and "R" is distance in meters.

The following modification to this formula (considering the speed of light) gives us the energy dilation of light coming from a gravitational body, such as a star, resulting from its own gravity:

$$Energy\ Dilation = \frac{Gm}{c^2r + Gm}$$

For fast moving bodies, we can add mass from kinetic energy to mass at rest and then plug it into this formula for gravity.

Kinetic energy may be found using Einstein's formula of

$$W = mc^2 \left(\frac{1}{\sqrt{1 - v^2/c^{2)}}} - 1 \right)$$

W = what Einstein used to represent kinetic energy.

Combining our dilation from gravity formula with Einstein's kinetic energy formula we arrive at a formula for computing dilation from the gravity of a moving body:

$$x = \frac{Gm}{Gm + c^2r \sqrt{1 - v^2/c^2}}$$

This tells us the reduction of energy (red-shift) and dimming of a star. To see it in action, let's imagine our sun moving at various speeds. Our sun's mass is 1.97×10^{30} kilograms. Its radius is 695,950,000 meters.

At various speeds we find the following dilations in this chart:

We can see that the dilation effect due to increased gravity resulting from speed is very mild and only becomes significant as it approaches light speed. Still, the increase does reach 100% at the speed of light.

This effect dims and reddens light emitted from a star. Because it occurs before the spectral lines are formed, we do not need to consider it when calculating red-shift of spectral lines. But, it should be considered when calculating distance of a very fast moving star based on brightness.

Speed	Dilation
0	2.100×10^{-6}
0.1c	2.110×10^{-6}
0.2c	2.140×10^{-6}
0.3c	2.201×10^{-6}
0.4c	2.291×10^{-6}
0.5c	2.425×10^{-6}
0.6c	2.625×10^{-6}
0.7c	2.941×10^{-6}
0.8c	3.500×10^{-6}
0.9c	4.816×10^{-6}
1.0c	1.000

(This effect may play a very small part in why stars in the most distant clusters of galaxies appear redder than their spectral classes indicate they should be.)

Chapter 8 - Energy Enhancement

As time and distance become less, the energy level of electromagnetic radiation increases. This radiation includes light, X-rays, radio, and more. It is pure energy. Massless.

It comes in the form of waves and at the speed of light. We have 3 ways of measuring it: wavelength, frequency of waves, and energy level, and they are all related through its constant speed.

Longer waves come less frequently than shorter waves. Each wave represents a quantity of energy. So, more frequent, or shorter, waves represent a higher energy level.

As we travel faster, we experience less time and distance.

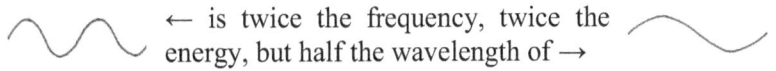 ← is twice the frequency, twice the energy, but half the wavelength of →

Any energy we receive then, would be experienced at a higher level.

This red-shift from dilation of the source will appear as shifted spectral lines. While spectral lines are created by the star's atmosphere, not the star, the atmosphere is traveling with the star experiencing the same dilation.

When we receive energy of a certain wavelength and experience it as less, the wavelength to us is effectively shortened. To go along with that, we receive a number of waves in a period of time. That time is also shortened as much as distance is. So, all energy we receive is at a shortened wavelength, increased frequency, and therefore higher energy.

All energy gets enhanced or "blue-shifted" to the degree of our distance and time dilation. This would not affect changes among objects traveling with us. Each object would experience emitting a normal level of energy, while actually emitting a lower level. And the object receiving it would experience it as being enhanced back to its normal level. This is similar to the way the Doppler effect gets canceled by traveling with the wave source.

Travel Speed	Emitted Energy Reduced By	Received Energy Increased By	Net Change In Energy
0.1c	0.5%	0.50%	0
0.2c	2.0%	2.04%	0
0.3c	4.6%	4.82%	0
0.4c	8.3%	9.05%	0
0.5c	13.4%	15.47%	0
0.6c	20.0%	25.00%	0
0.7c	28.6%	40.05%	0
0.8c	40.0%	66.67%	0
0.9c	56.4%	129.36%	0
1.0c	100%	Infinity	(Cannot experience anything)

To an outsider though, assuming he's not in motion and is instead experiencing true energy levels, he would find our energy reduced. A bystander who's at rest in space would notice that a body in motion emits energy at a lower level. This coincides with also noting that the moving body experiences less time and distance.

Effect On Our View of The Universe

This concept, applied to an expanding universe, would mean that all observers moving away from the center could look toward the center of the universe and see energy at normal levels, not appearing as red-shifted because the observer is the one in motion. He would receive energy from non-moving or slower moving objects as enhanced. (This works against, and is lesser than, the Doppler effect which would create a red-shift in the same situation.)

He could also look toward the outer universe where bodies are moving away faster than he is. Since those bodies are moving faster, they experience less time and distance. And, while to them they're producing a normal level of energy, they're actually producing a lower level than even the observer is. Even with his enhanced perception of energy, those bodies would seem to be at a lower level (red-shifted). While he lives in slow motion, they live in slower motion. This would be added to the Doppler red-shift in the same situation.

This is a relativistic effect which is independent of, but would be added to or subtracted from, the Doppler effect to create altered red-shifts and blue-shifts. It is based on the dilation effect on the body being observed compared to any dilation which the observer has, just as we compared time dilation between the flight clock and the Earth clock.

An example of how this would add to a red-shifted galaxy is as follows.

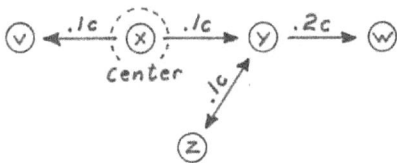

("c" = the speed of light and every object is moving away from their neighbors at 10% of "c", or 0.1c.)

- "y" sees "x" as red-shifted by the Doppler effect by 0.1, but enhanced by dilation by 0.005 due to "y" speed of 0.1c.
- "y" sees "z" as red-shifted by the Doppler effect by 0.1 only. There is no effective dilation because they both have the same absolute speed of 0.1c and so their dilation is equal.
- "y" sees "w" as red-shifted by the Doppler effect by 0.1 and more dilated than himself. (0.02 - 0.005 = 0.015 dilation) "w" would appear red-shifted by 0.115 total.
- "y" sees "v" red-shifted by the Doppler effect by 0.2. Their dilation is equal.
- "y" would see the greatest energy coming from the center of expansion, the least energy furthest from center.

This combination of energy dilation and the Doppler effect creates difficulty in determining speeds of objects in our distant universe. Objects moving toward us or sideways to us could deceive us into misinterpreting their speed due to their altered color shift.

For purposes of evaluating distant galaxies however, if we assume that most are simply moving away from us, we can apply the following formula to calculate their total red-shift.

$$1 - \left(1 - \frac{v}{c}\right) \sqrt{1 - \frac{v^2}{c^2}}$$

Energy enhancement is a known effect of relativity. My suggestion is that it be applied to the whole universe in recognition of absolute movement.

Chapter 9 – Movement of Energy

So far, we've considered the absoluteness of time and distance and how it affects movement. There's one more aspect of movement which we need to consider, and it too is absolute.

Velocity is speed (distance/time) and direction. In absolute space and time, direction is viewed as absolute as well. And, while we might believe that direction is not an issue because it's not involved in the relativistic effects, it is absolute and is involved in movement. Direction does make a difference in the velocity of anything which moves only according to space. That would be energy, the one thing which is relative to space only. It does not move relative to any other reference frame.

The movement of energy (such as light) was a great mystery for thousands of years. But, in 1865 Maxwell explained that it moves itself. With his 4 famous equations, he laid out how energy propagates as a back and forth of electric fields creating magnetic fields creating electric fields. In this way, we can imagine that light crawls from place to place on its own.

In 1905, Einstein's special theory of relativity showed us how the movement of energy is relative to each observer. Or, at least its experienced speed is. It's actual speed and direction, I suggest, is not.

Einstein wrote that, "Light always propagates in empty space with a definite velocity "v" that is independent of the state of motion of the emitting body." (We had not yet begun using "c" to represent the speed of light.)

If motion was only relative between 2 bodies, if there was no absolute motion, then how could light's motion be independent of the emitting body? If all motion is only relative, then for light to have any motion at all it would have to be related to its emitting body before all else. I will show here that light's speed and direction are independent of its source and are relative to space itself.

Light's Independence

The velocity (speed and direction) of light is not affected by the movement of its source. We know this because of the Doppler effect.

If light had inertia due to the movement of its source, the inertia would cancel the Doppler effect and alter the speed of light.

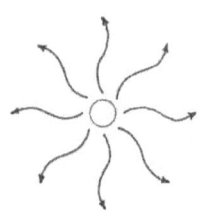

Imagine a star at rest in space. It emits light equally in all directions and that light travels at the same speed and is of the same wavelength in all directions.

Now suppose that star is moving through space at 1/2 the speed of light. Since its photons can only travel at the speed of light, their waves get shorter by 50% in the direction it's going.

The star emits the same number of waves per second in all directions, but they can't get as far away from their source in the direction of movement. And, behind the star the waves are stretched because the star has traveled as each wave was emitted, which makes each of them longer. This is the Doppler effect.

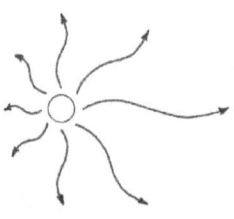

If the light emitted from the star carried inertia due to the star's movement, its wavelengths would remain the same as emitted according to bystanders. The reason is because the light would travel 50% faster than the speed of light in front of the star and 50% slower than the speed of light behind it, according to bystanders.

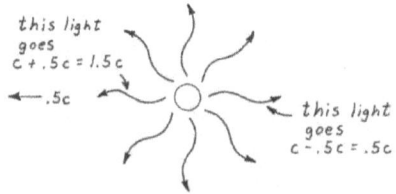

A star moving toward us at 1/2 the speed of light would not appear to have 50% shorter waves, because the photons would be traveling 50% faster and they would be of the same wavelength as emitted.

Photons from a star moving away from us at 1/2 the speed of light would be of the same wavelength because they would be traveling at half the speed of light.

Since we experience the Doppler effect as a source of light moves toward or away from us instead of a change in the speed of light in direct proportion to its movement toward or away from us, we know light moves independent of its source. (Doppler's blue-shift is the more reliable evidence, since red-shift may also, as I propose, be a result of energy dilation.)

Direction

Consider aiming light at a target. If light doesn't move according to its source, then we don't need to consider any momentum it may get from it. This applies in any direction.

Suppose I'm flying through space on my rocket. And, I'm going very fast. (I know this because I'm burning a lot of fuel.) I see a meteoroid and fire a laser pulse at it as I pass. Because I fired straight at it, and I'm a pretty good shot, I know I'll hit it. Here's why.

Once the light pulse leaves my gun it travels just the way I aimed it. Light isn't affected by my speed. It propels itself toward the meteoroid on its own, at the speed of light.

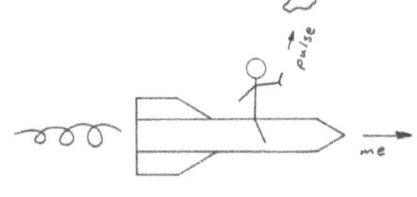

This is not like an old-fashioned gunpowder gun. Shooting a bullet from this type of gun is much more difficult. The reason is because the bullet in my gun is already traveling very fast along with me before I fire it. When I fire, then, it has motion going forward with me plus the motion the gunpowder gives it to go to the side in the direction I aimed.

The bullet, having forward and sideways motion, ends up traveling at an angle and even faster than either of those speeds.

It's calculated as *rocket speed* ^{2} + *gunpowder speed* ^{2} = *bullet speed* ^{2} which is why bombing from an aircraft is so difficult. The projectile carries with it the inertia of the object firing it. So how do we know my laser pulse won't do the same thing?

Two reasons:

- If it did, it would end up traveling faster than light. (*light speed* ^{2} + *rocket speed* ^{2} = *pulse speed* ^{2})

- Secondly, the light pulse, unlike my bullet, didn't even exist before I fired it. So it can't have any forward momentum.

Photons are not a part of matter, but are emitted by it. The light pulse was emitted by my laser, but not propelled by it. It took off on its own in relation to space itself. From the real location of its creation in the real direction it was aimed.

This can tell us about absolute movement. For example, if the meteoroid is not moving I'll definitely hit it with my laser. Because I'm that good. (You noticed the cowboy hat didn't you?) But, if the meteoroid is moving fast enough I'll miss it.

Here's how that would work. For simplicity, let's go back to before I launched my rocket. I'm standing still with my rocket and see a meteoroid fly by. I fire my laser pulse at it, but the pulse goes straight, as it always does, and the meteoroid gets out of the way just in time. It was moving very fast and was far enough away that my laser pulse took a little while to get there and I missed.

We can apply this principle to prove that there is absolute movement and to determine what our absolute movement is.

Determining Absolute Movement

Let's go back into space and see how the movement of light tells us about our own movement. We'll return to our previous scenario in which one of us is moving but we can't tell whom.

Remember that I see you approach from my left and leave to my right.

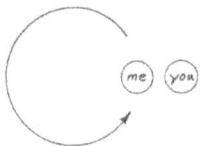

And, you see me approach from your left and leave to your right.

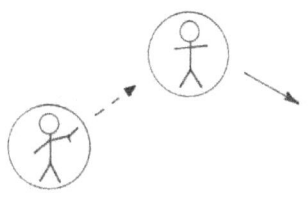

One of us seems to be orbiting an invisible mass so that we both feel weightless. Instead of comparing our watches, let's see where a pulse of light goes.

As we pass each other, I aim my laser directly at you and fire a pulse. If movement is only relative to the observer, then I would see my pulse miss you, because to me you're moving very fast and get out of the way before being hit.

But to you I'm the one moving. Since you're at rest, the laser pulse hits you as I move away from my firing position.

If that laser pulse is strong enough, you could be dead to you, but alive to me.

75

We don't find these differences in reality. So, according to our one reality I missed you. And, that's not because I don't know how to aim a laser. It's because you moved out of the way. You're the one who's actually in motion.

This tells us about our absolute movement since that's the type of movement light has. Because light moves relative to space itself it can be used to detect the movement of anything else relative to space.

On Earth we don't find problems because light seems to go straight here. We don't find lasers not hitting their intended target. That's because our target is nearly at rest relative to space. But, try hitting a target on the moon with a laser and you'll have to aim ahead of it. You'll have to aim for where the moon will be when your laser pulse gets there. The moon is traveling 1 km/sec in its orbit and it takes light 1.3 seconds to get to it from here.

Computer laser disk readers do not perform differently when we orient them in different directions. The laser isn't missing its target. This really only tells us that any movement we have through space isn't enough to noticeably disrupt our aim of light.

To determine our absolute movement, we could use light's movement to tell us. Fire a laser in various directions and measure if its destination varies at all. This will have to be done with extreme precision to detect our movements which seem to be very slow compared to light. So we'll need to amplify our results. Here's one way you can do it yourself.

The Cosmic Speedometer

Set up a laser nearly perpendicular to 2 parallel mirrors facing each other, all on a portable platform, so that its beam bounces back and forth between them until it finally reaches the end of the mirror where a ruler is. (You'll want the angle of the laser to be easily and

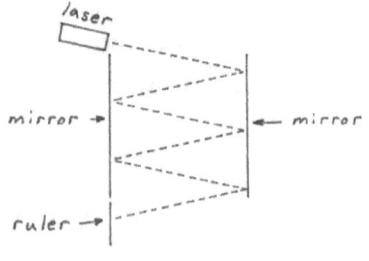

precisely adjustable to fine degrees, such as with a thumbscrew.) The more times you can bounce the light, the more distance you can replicate (the longer your beam will effectively be).

If the distance between your mirrors is a foot, and you can bounce the beam back and forth 100 times, then you've effectively created a beam of light of 100 feet (actually a little longer, but this is not a significant amount).

The longer you can make the beam, the better you'll be able to detect slower speeds. Without the mirrors, you could detect for speed differences in the 10's of millions of miles per hour. With the mirrors, you can detect much slower speeds.

The calculations are simple. The deviation of your beam, compared to the length of the beam, is your speed as a portion of light speed. So, if you're traveling perpendicular to your beam at the speed of light, the beam would appear to go at an angle like this.

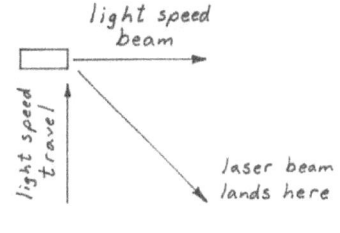

While the light travels 1 foot across your device, your device moves 1 foot through space.

Our solar system travels around the galaxy at about 250 km/sec. That's about 1/1200 of the speed of light. That means that if your mirrors are 1 foot apart, and you bounced your beam between them 6 times (for 12 segments of light), the end of that 12 foot beam would deviate by 1/100 of a foot from where it should land.

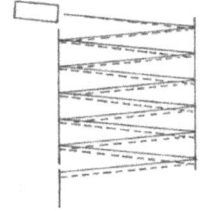

1/1200 x 12 = 1/100

You could rotate your device and find the beam moving between 1/100 of a foot in 1 direction and 1/100 of a foot in the other direction. That's about 1/8 inch deviation each way.

That's the basic concept. There are a few more factors to consider, though.

Design Details

This design is limited by the angle of the beam, which means it can only detect within a narrow range of speeds with each setup. So, it needs to be calibrated for each detection it performs.

The range of speeds a particular setup will be able to detect is determined by the number of bounces of the beam, distance between mirrors, and length of mirrors.

Here are the formulas:

(Units of measure and units of speed are your choice, but they must be consistent.)

- Speed, which a unit of distance represents is:

$$\frac{(unit\ of\ measure) \cdot (speed\ of\ light)}{(2\ segments\ per\ bounce) \cdot (no.\ of\ bounces) \cdot (distance\ between\ mirrors)}$$

- The variation (as a portion of light speed) you'll be able to detect will be:

$$\frac{length\ of\ mirrors}{(number\ of\ bounces) \cdot (2\ segments\ per\ bounce)}$$

For example, suppose you use 2 mirrors 14 inches long and placed them 12 inches apart. (I'm using these numbers just to show that you may use any dimensions you'd like.) 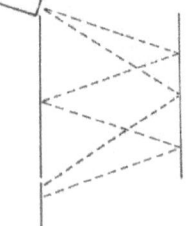 You aim the beam at the opposing mirror so it bounces once and lands on your ruler just beyond the laser-side mirror (1 bounce, 2 segments of light). Then you slowly draw the beam back toward straight while you count bounces each time your laser point disappears from your ruler and re-appears.

That double segment of light, as you slowly straighten the laser, turns into 4, then 6, then 8 segments.

Once you've counted 25 bounces, you may stop and use the formulas to determine what your device is calibrated to detect.

$$\frac{(\frac{1}{16} \ inch) \ \cdot (670 \ million \ mph)}{(2 \ segments \ per \ bounce) \ \cdot (25 \ bounces) \ \cdot (\frac{192}{16} \ inches \ between \ mirrors)}$$

$$= 70,000 \ mph \ per \frac{1}{16} \ inch \ on \ your \ ruler$$

Since 14 inches is $\frac{224}{16}$ inches, your range with this setup is limited to:

$$\frac{\frac{224}{16} \ inches}{(25 \ bounces) \cdot (2 \ segments \ per \ bounce)} = \frac{4.5}{16} \ inch$$

When using the device, you'll want to rotate it all the way around. This setup will allow you to measure 70,000 mph or 140,000 mph in either direction. It would work fine for detecting Earth's orbit around the sun.

Calibrate and Measure

Every time you perform an experiment, you'll be determining relative speed and direction and will need to calibrate, or set up your device, for that measurement.

To begin, set your laser beam to land on the middle of the available marks on your ruler, just past the mirror. Remember you're only using the first few marks of your ruler as the formula indicates.

Then, rotate the device in any direction slowly and watch for the beam to move. If the beam moves off the ruler to the mirror or disappears from your ruler and reappears then your setup is too sensitive. You'll need to restart and use less bounces.

If the beam doesn't move, then you're not traveling that fast. Try detecting for a slower speed. If the beam moves, keep rotating your device until it reaches its greatest extent. Then, continue to rotate until it reaches its greatest extent the other direction. The difference between these 2 extremes is twice your actual speed. When oriented to one extreme, your speed is the deviation from the midpoint between the extremes. And your direction is perpendicular to your laser beam, or parallel to your mirrors and ruler in the direction opposite of the laser's deviation.

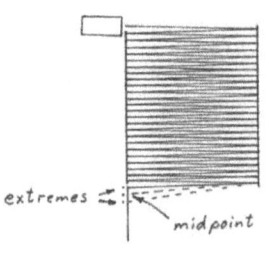

So, in the setup we just discussed, you have an available range of 4 1/2 sixteenths of an inch. If you reach one extreme and continue to rotate and reach the other extreme in the opposite direction, and those 2 extremes are 2 sixteenths of an inch apart, then you have a deviation of 1 sixteenth in each direction. You're detecting 70,000 mph.

If you're uncertain of your movement through space before beginning to test for speed, start with a reduced sensitivity, detecting for fast movement first, then work your way toward detecting lower speeds. Here are some speeds you might like to detect for:

- Speed of Earth through galaxy: 450,000 mph
- Speed of Earth around the sun: 66,900 mph
- Speed of Earth revolving daily at various latitudes:

0 degrees (equator):	1038 mph
10 degrees:	1021 mph
20 degrees:	975 mph
30 degrees:	898 mph
40 degrees:	795 mph
50 degrees:	667 mph
60 degrees:	519 mph

Using longer mirrors or placing them further apart will help you increase the effective length of your laser beam. This will make more precise and slower speed measurements.

And, remember that what you're detecting is absolute movement through space, which may consist of multiple combined motions. It may be in any direction, including up or down.

Effect On Traveling Systems

When a star emits light, the light travels straight, according to space, from the location of the star when the light was emitted. This tells the observer of the location of the star at the time the light was emitted. If the star is moving in any direction other than directly toward or away from the observer, we'll see it move slowly and time delayed by its distance.

If the observer is not moving significantly, then a star's speed and distance in relation to the observer will determine its brightness and color shift.

The color shift as a result of the Doppler effect, and also as a result of energy dilation, will each cancel out if the observer has the same movement as the star. This means that a system of bodies traveling together will not notice its own color shifts. It will not notice its own movement if it's considering only color shift. This is because of light's absolute movement through space regardless of any other movements.

But, while light's absolute movement through space causes a color shift noticeable only to observers outside of the traveling system, it also creates an effect noticeable only to observers within the system or traveling in some way.

This effect is a directional energy shift. The light will seem to have an altered direction to observers traveling with it, but not to outsiders.

Directional Energy Shift

When an observer is moving, photons seem to be redirected. Their direction of travel seems to be altered and their destination shifts in the direction opposite of travel. This is a phenomenon created by the absolute movement of the observer. It shows itself most simply inside a moving system.

In a system of objects traveling together, the result of their travel is that the energy they exchange will shift destination. It doesn't change direction. The energy moves straight and in the direction it was emitted, according to space. But as the observer travels, he moves into or out of the path of the photons.

This is the cosmic speedometer as it naturally occurs. It's a phenomenon experienced by every traveling energy system. Computers, biological systems, solar systems, and galaxies all exchange energy internally in the form of photons. And, as those systems move in one direction, the energy exchanged gets shifted in the opposite direction.

The reason for this is that pure energy (such as photons) travel according to space, regardless of the movement of its source. And, although it goes very fast, it always takes some amount of time to get to a destination. If that destination is moving in any way, exactly where that energy is received can change.

Just as we saw in the movement of light, the amount of shift is equal to the speed of the system as a portion of light speed, while moving perpendicular to the direction of the energy.

For all other angles, the shift is equal to the sine of the angle of incidence multiplied by the speed of the moving system as a portion of light speed.

$$Energy\ Shift\ =\ v/c$$

(v/c = the speed of a system expressed as a portion of light speed)

or

$$Energy\ Shift\ =\ v/c\ \sin \phi$$

(ϕ = the angle of incidence between energy direction and direction of the system)

(An adjustment for angle of incidence may be added to any of the formulas to account for movement at angles other than directly toward or away. I haven't been adding it because I've been generally assuming objects are moving directly toward or away from each other. But, in the case of energy shift, it's easy to see how this could have important consequences for small portable systems such as electronics and living creatures which have complicated movements and changing orientation.)

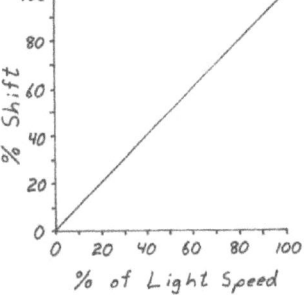

At rest   Moving

As with relativistic effects, it is generally not a concern at the speeds we normally travel. However, unlike dilation of time and distance, this effect is not exponential. It's linear.

Energy shift increases at a constant rate through all speeds. This means that while we do have to go fast to notice it, we don't have to approach light speed to see significant effects.

Effects of Energy Shift

In rare circumstances energy is directed precisely at its intended destination, such as with a cosmic speedometer. If precision is required, any shift of that beam, or photon's, destination could disrupt the operation of that system. A laser disk reader could fail to perform while the device is moving at a high speed perpendicular to its beam. A land surveyor could miscalculate measurements if he's using lasers aimed at precise angles.

Another good example would be a star as observed while traveling with it. Photons emitted by a star go every direction. A star moving through space would appear to have a shift of intensity in the direction opposite its movement, for any observers traveling with it. The amount of the shift would be equal to the speed of the star as a portion of light speed, adjusted for viewing angle.

From directly in front of, or behind the source, brightness will not be altered dramatically by this shift because the apparent angle remains nearly the same. However, viewing the source from the side, the observer will notice the greatest shift. And because of this, the light will be more concentrated or less concentrated at various angles. The more the observer's view of the energy source gets larger, the dimmer it will appear. As the source appears smaller the rays will concentrate and effectively look brighter. We'll explore this in more detail as a visual distortion.

This would not affect the brightness of stars as they move toward or away from an observer at rest. It's not about actual energy emission, but the effect on an observer moving into or out of energy's path. So it would cause a shift based on the observer moving with a moving source as part of a system, or just having some amount of absolute movement.

Gravitational Shift

Just as light shifts according to observers within a traveling system, gravity should too. Regardless of the mechanism which makes gravity work, whether its energy is represented by a particle or a warping of spacetime, it behaves in such a way as to make it subject to energy shift.

We know that gravitational waves travel at the speed of light, as evidenced by data collected from gravitational wave detectors, such as the Laser Interferometer Gravitational-Wave Observatory (LIGO).

LIGO has locations at Hanford, Washington and Livingston Parish, Louisiana. In 2016, its detectors recorded identical signals, but received them 7 milliseconds apart at the two locations. They detected the same wave which was traveling at light speed.

That tells us that gravitational waves move independently of their source. Light speed is fixed. To move at a fixed speed, that speed cannot depend on its source. Otherwise, it would be affected by the source's speed. Additionally, to move at light speed they could have no mass, so they couldn't carry inertia from their source anyway. Gravitational waves must move according to space itself, as light does. And, therefore, be subject to the same shift as light.

The effect of gravitational shift, like energy shift, would most simply occur within a traveling system. Like the apparent light coming from a star, gravity would effectively be stronger as it's concentrated at certain angles of a moving body and weaker at others.

In a traveling gravitational system, such as a solar system or galaxy, the enhanced or reduced gravity at various angles of each body in the system would partially cancel each other. This is because their relationship with each other is opposite. For example, if a star is traveling with a planet in a way that the planet is not directly beside it but is 20° to its rear, then to the planet the star is 20° to its front. We'll see how this may be calculated using the same formula as visual effects.

These proportional changes work against each other because gravitational forces get multiplied together, not added. The shift of light does not cancel because light is one way. It is sent and received. Gravity is mutual.

Visual Distortion

Let's consider the view a traveling observer has because this tells us of all the energy he receives, not just visible light. It tells us what he sees and what's affecting him.

The shift of energy changes where things appear to be for the observer because perception of location is based on the angle of the light received.

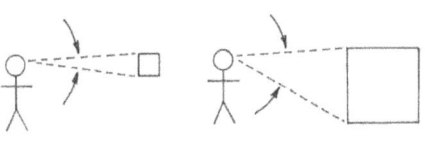

Angle also determines the size objects appear to be because that is based on how great the angle is between light coming from one end of an object and light coming from the other.

The orientation that objects appear to be positioned at is created by angles. It's based on the view of the object carried by the light we run into as we travel.

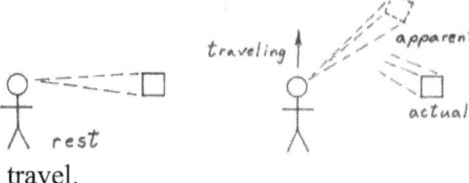

The view we run into changes with speed of travel. As an observer moves, objects in the direction of travel appear slightly narrower or compressed. To his sides, even narrower and shifted toward the front. And toward the back, objects look larger as they stretch toward the front. But directly to the rear are near normal size. They will appear turned toward the observer too, by their amount of location shift. The change occurs most dramatically in the side view. Stars would appear arranged like this to a fast-moving observer. They would appear to gather in front of him in his quarter view.

Any point directly in front of or behind a traveling observer will still be there. This is because its apparent angle doesn't shift, but all else will look swept forward according to travel speed compared to light speed, and the viewing angle compared to travel direction.

An easy way to visualize this is as a traveling observer looking to the side. The change in angle that a point appears to be located at may be calculated using the formula of:

$$\theta_1 = tan^{-1}(tan\ \theta_0 + v/c)$$

θ_0 = the actual angle in relation to the observer's perpendicular view when the light was emitted.

θ_1 = the angle the point appears to be located at.

86

From our view, looking perpendicular to travel, our angle is 0°. Looking toward the direction of travel, we'll measure a positive angle. To the rear we'll use a negative angle. If distances are used, this formula applies:

$$\theta_1 = tan^{-1}(y/x + v/c)$$

y = distance in the forward direction
x = distance in the sideways direction

The table on the following page shows the direction that points appear to be located at, based on their actual position and the observer's speed.

We can see from the table, that an object which should span from the 20.0° to the 30.0° directions of our vision (10.0° total) will only appear to span from the 29.4° to the 37.9° directions (8.5° total) while we travel at 20.0% of light speed. Its apparent location has shifted toward our front by about 8.7°. Its orientation has rotated toward us by the same amount. And its width has decreased by 15%.

87

Actual Angle	Apparent Angle				
	at 0.2c	at 0.4c	at 0.6c	at 0.8c	at 1.0c
90°	90.0°	90.0°	90.0°	90.0°	90.0°
80°	80.3°	80.6°	80.9°	81.2°	81.5°
70°	71.3°	72.4°	73.4°	74.3°	75.1°
60°	62.6°	64.9°	66.8°	68.4°	69.9°
50°	54.3°	57.9°	60.8°	63.3°	65.5°
40°	46.1°	51.1°	55.2°	58.6°	61.5°
30°	37.9°	44.3°	49.7°	54.0°	57.6°
20°	29.4°	37.4°	43.9°	49.3°	53.8°
10°	20.6°	30.0°	37.8°	44.3°	49.6°
0	11.3°	21.8°	31.0°	38.7°	45.0°
-10°	1.3°	12.3°	22.9°	31.9°	39.5°
-20°	-9.3°	2.1°	13.3°	23.6°	32.5°
-30°	-20.7°	-10.1°	1.3°	12.6°	22.9°
-40°	-32.6°	-23.7°	-13.5°	-2.3°	9.1°
-50°	-44.7°	-38.4°	-30.6°	-21.4°	-10.8°
-60°	-56.9°	-53.1°	-48.5°	-43.0°	-36.2°
-70°	-68.6°	-66.9°	-65.0°	-62.8°	-60.2°
-80°	-79.6°	-79.2°	-78.8°	-78.4°	-77.9°
-90°	-90.0°	-90.0°	-90.0°	-90.0°	-90.0°

These changes are shown on the table on the next page. Enlargement is in percentage. Direction and rotation are the same and are shown in degrees.

Enlargement, Direction, and Rotation

Actual Angle	at 0.2c	at 0.4c	at 0.6c	at 0.8c	at 1.0c
90°	0.0%|0.0°	0.0%|0.0°	0.0%|0.0°	0.0%|0.0°	0.0%|0.0°
80°	-6.5%|+0.3°	-12.0%|+0.6°	-17.0%|+0.9°	-21.5%|+1.2°	-25.5%|+1.5°
70°	-11.5%|+1.3°	-21.5%|+2.4°	-29.5%|+3.4°	-36.0%|+4.3°	-42.0%|+5.1°
60°	-15.0%|+2.6°	-27.5%|+4.9°	-37.0%|+6.8°	-45.0%|+8.4°	-52.0%|+9.9°
50°	-17.5%|+4.3°	-31.0%|+7.9°	-42.0%|+10.8°	-51.0%|+13.3°	-58.0%|+15.5°
40°	-18.0%|+6.1°	-32.0%|+11.1°	-44.5%|+15.2°	-53.5%|+18.6°	-60.5%|+21.5°
30°	-16.5%|+7.9°	-31.5%|+14.3°	-43.5%|+19.7°	-53.5%|+24.0°	-61.5%|+27.6°
20°	-13.5%|+9.4°	-28.5%|+17.4°	-40.5%|+23.9°	-51.5%|+29.3°	-60.0%|+33.8°
10°	-9.5%|+10.6°	-22.0%|+20.0°	-35.5%|+27.8°	-47.0%|+34.3°	-56.0%|+39.6°
0°	-3.5%|+11.3°	-11.5%|+21.8°	-25.5%|+31.0°	-38.0%|+38.7°	-49.5%|+45.0°
-10°	+3.0%|+11.3°	-1.5%|+22.3°	-11.5%|+32.9°	-24.5%|+41.9°	-37.5%|+49.5°
-20°	+10.0%|+10.7°	+12.0%|+22.1°	+8.0%|+33.3°	-3.5%|+43.6°	-17.0%|+52.5°
-30°	+16.5%|+9.3°	+29.0%|+19.9°	+34.0%|+31.3°	+29.5%|+42.6°	+17.0%|+52.9°
-40°	+20.0%|+7.4°	+41.5%|+16.3°	+59.5%|+26.5°	+70.0%|+37.7°	+68.5%|+49.1°
-50°	+21.5%|+5.3°	+47.0%|+11.6°	+75.0%|+19.4°	+103.5%|+28.6°	+126.5%|+39.2°
-60°	+19.5%|+3.1°	+42.5%|+6.9°	+72.0%|+11.5°	+107.0%|+17.0°	+147.0%|+23.8°
-70°	+13.5%|+1.4°	+30.5%|+3.1°	+51.5%|+5.0°	+77.0%|+7.2°	+108.5%|+9.8°
-80°	+7.0%|+0.4°	+15.5%|+0.8°	+25.0%|+1.2°	+36.0%|+1.6°	+49.0%|+2.1°
-90°	+0.0%|0.0°	+0.0%|0.0°	0.0%|0.0°	0.0%|0.0°	0.0%|0.0°

In this table we see that if we were traveling at 40% of light speed an object which is actually 10° toward our front will appear to be an extra 20.0° forward, 22.0% smaller, and rotated 20.0° toward us, so we see more of its front than expected for that viewing angle. And if we were traveling at 80% of the speed of light, an object which is actually 50° to our rear would appear 28.6° further forward than normal, rotated toward us by 28.6° so we see more of its front side. And it would appear 103.5% bigger than it really is.

This applies to objects a viewer is traveling with or past, the inside of a spacecraft as an astronaut looks around or the sky outside him as he looks out a window.

The reason an observer in a traveling solar system would see its sun as brighter while behind it, is because as the angle of the light becomes smaller, his eyes will collect more photons than usual per degree of vision. While in front of his sun, he would look back and see it as larger, occupying more of his view and the light would be less concentrated. He would receive less photons per degree of vision.

If the observer wasn't traveling, only the sun was moving, he would see it as normal in its location as it passes by. No shift of energy. The location he sees it in would just be delayed by the time the light takes to get to him.

It's the observer's travel which changes how things appear, and how the energy is received. Moving objects simply appear in their old position but still normal in their size and orientation to a viewer at rest.

Energy Migration

As energy shifts it may transfer simply, like light emitted by a star and received by an observer on a planet traveling with that star.

But, in a system which has a flow of energy, there is a series of transfers. A sort of enhancement of energy, or reduced resistance, will be created in the direction of shift (opposite that of travel), and a resistance will be created in the direction of travel.

Most energetic systems move energy (exchange photons) in a zig-zag type of progression, because photons are generally emitted in a seemingly random direction. This is referred to as "spontaneous" emission and was described by Einstein in 1917. Only laser, "stimulated" emission, is known to have a predetermined direction. This occurs in most all energy pathways because, at the molecular scale, they are not narrow. They are many molecules wide.

Photons are sent in all directions, but their activity goes with the direction of energy flow.

The effect of directional shift on these systems is a general accumulation at one end of the system and reduced activity at the other.

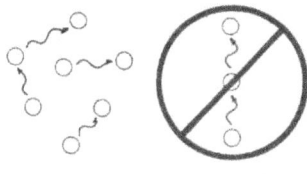

Like this   *Not like this*

Energy shift is able to create a migration of energy within a system of many particles or objects.

The cores of stars are not dramatically affected by this because their particles are generating their own emissions. They are not so much transmitting energy. The outer layers of a star, however, are transmitting. They are passing the energy generated by the core outward. Moving stars, then, will experience energy migration in their outer layers. This will make all moving stars, viewed by all observers, appear brighter on their backside and dimmer on their front, because they will actually be brighter and dimmer.

The thickness of the transmitting layer of stars, like the thickness of electrical pathways does not affect the degree of migration or the effective change in resistance. (The size of a pathway affects total resistance, but not a degree of change of resistance.)

Stars moving away from us will not look so far away. Stars moving toward us will be closer than they appear. And the extent of this effect will be the average of shifts in every direction, weighted according to their circumference.

$$Migration = \frac{\sum_{\theta=-90}^{\theta=+90} 2\pi \cos \theta \cdot \sin\left[tan^{-1}\left(tan\, \theta + \frac{v}{c}\right)\right]}{\sum_{\theta=-90}^{\theta=+90} 2\pi \cos \theta}$$

(This amount may be estimated, since it's range is only from 60% to 52% of v/c as speed goes from rest to light speed.)

Effect On Electrical Systems

Migration only applies to objects such as photons which are exchanged and whose movement is in relation to space only, not related to the movement of the system that its objects are being exchanged in. It cannot apply to objects which carry inertia like electrons. Objects with inertia move relative to their system. Because they have mass, their behavior can be predicted by general relativity. Energy (objects without mass) can be predicted by absolute relativity since it moves independently of its system.

Electricity is the flow of electrons, which do not move independently of the system they're a part of. They have mass and carry the inertia of the moving system with them. So, they do not experience energy migration directly. But, electrons get energy to move as they absorb photons. And, photons do experience energy migration directly. In this way, electrons should experience energy migration indirectly as a result of their photon exchange.

The effect of energy migration on electron flow should be equal to the effect the migration has on photons. In the same proportion, in the same direction.

The result would be an increase or decrease of resistance within an electrical system, depending on the system's speed and direction and the direction the electrons are flowing.

We can add this feature to Ohm's law for calculating voltage using current and resistance.

Ohm's law is: $V = IR$

(V = voltage, I = current, R = resistance)

So, for a traveling system the formula is:

$$V = \frac{IR}{\cos \emptyset \left[1 - \frac{\sum_{\theta=-90}^{\theta=+90} 2\pi \cos \theta \cdot \sin \left(\tan^{-1} \left(\tan \theta + \frac{v}{c} \right) \right)}{\sum_{\theta=-90}^{\theta=+90} 2\pi \cos \theta} \right]}$$

v = speed of system
c = speed of light
ϕ = the angle of electrical flow compared to the
direction of movement of the system

"v" is positive if it's in the direction of electrical flow. It's negative if it's in the opposite direction of electrical flow.

This allows a system at rest to experience Ohm's law as it currently is. And, any movement through space adjusts the results.

Approaching light speed in the direction of electrical flow, resistance will grow to infinity.

Approaching light speed in the direction opposite of electrical flow, since velocity then is negative, resistance approaches zero.

This resistance caused by energy migration would apply similarly to any system which operates based on photon exchange, which includes all chemical reactions, including biological. And it will alter the performance of these systems according to their speed, eventually reaching their operable limit. If a system involves energy flowing various directions, it may be unable to function at any orientation at a high enough speed.

Gravity Migration

We have evidence that gravity is not instantaneous but instead travels at the speed of light. Moving at light speed means it can have no mass. It also means it cannot move according to its source and must move according to space, as light does. We also know that the emission of gravity is a decay of matter.

Because of that, if the first law of thermodynamics, the law of conservation of matter and energy, is to be true, then gravity must be a form of matter and/or energy. And it would have to be the latter to move at light speed.

93

As a form of energy, it must increase the total energy of any body that it may be absorbed by, just like its emission decreases the total mass/energy of the body that emitted it.

Energy will migrate if it's absorbed and re-emitted, changing the total mass/energy of each object it interacts with.

We know that gravity is emitted since it travels independently of its source. We know it takes some of the energy of its source with it. So is it absorbed?

There are 4 known forces in our world: gravity, the electromagnetic force, the weak nuclear force, and the strong nuclear force. At least three are represented by a particle and work by exchanging those particles.

- The electromagnetic force is carried by the photon.
- The weak nuclear force is carried by the W and Z bosons.
- The strong nuclear force is carried by the gluon.

Many scientists believe gravity is carried by the graviton. If it is, then perhaps it's exchanged, too.

These particles have all been found to exist, except for the graviton which has yet to be detected. But, being much weaker than the others, it should also be much more difficult to detect.

Gluons, W and Z bosons, and photons are all emitted and absorbed. They have energy (or mass) which they take from their source and contribute to their recipient.

Because gravity moves independently, I suggest it should be considered to be its own object, and described as a particle just like the other forces are. So, graviton seems to be the correct description of it.

And, since it is emitted, carries energy with it from its source, is capable of doing work, and moves at the speed of light, then it seems to be a particle of energy, like a photon or gluon is. (W and Z bosons have mass, and while they perform a task, they don't push or pull like photons and gluons, or gravitons do.)

Gravitons pull objects together. So do gluons and photons. They all are emitted, travel at the speed of light, carry energy from their source, can do work, and are a form of energy. To accomplish their pulling, gluons and photons are absorbed by the other object in the pulling scenario. I suggest gravitons are, too.

If this is the way gravity works, then gravity is subject to energy migration.

In a traveling system, such as a solar system, the effect would cancel. The sun's migration of gravity in the same direction as that of its planets would result in no change in their total gravitational attraction.

And, like energy migration affecting the real output (strength) of gravity of that star or planet, just as much as it is brighter on its backside and dimmer on its front side, it also has stronger gravity on its backside and weaker on its front.

Regarding other effects, remember that the Doppler effect cancels itself between objects moving together. Dilation does also, so a system will still function normally, but does leave it in a "slow" existence. Migration remains to affect the functioning of a system.

The Doppler Effect

On Brightness of Light

Just like an energy source emits waves at a rate, as parts of a photon (there are many waves in a photon), the photons, too, are emitted at a rate. The rate of waves (frequency) is its color, or energy level. The rate of photons is its brightness. As the frequency of a wave increases or decreases according to the Doppler effect, so does brightness.

While a source and observer are moving toward each other, and the observer receives a higher frequency, the observer also receives a higher brightness. Like the Doppler effect on color, the amount of change is directly related to speed and is calculated as its portion of the speed of light.

If the source and observer are moving toward each other at 7% of light speed, then the observer will see the light as 7% shorter in wavelength (bluer) and also brighter by 7% just as sound becomes higher pitched and louder when the source and recipient are moving toward each other.

If both observer and source moved away from each other at 22% of light speed, the observer would receive light which is 22% lower frequency (redder) and 22% less bright (dimmer).

This effect is the same regardless of which is moving.

Because it is in direct proportion to the rate the source and observer are moving toward or away from each other, it would counter and exceed the real brightness created by energy migration for an observer who is at rest in space.

For an observer at rest in space, he may misjudge the true distance of stars moving away from him by the amount which the Doppler effect of dimming exceeds the brightening caused by migration. Receding stars are actually closer than they appear.

On Gravity

The Doppler effect occurs when a wave source and recipient move at different velocities in relation to the waves which move at a fixed rate and independent of either. Sound moves at a fixed rate in relation to the air. Light moves at a fixed rate in relation to space.

Because gravity moves at a fixed rate independent of its source and recipient, it should display the Doppler effect.

Gravity between two bodies moving toward each other should be enhanced in direct proportion to their relative speed. And, moving away, their gravity should be reduced according to their relative speed. Two bodies a fixed distance from each other would experience normal gravitational attraction.

This would cause bodies to more easily move toward or away from each other than remain fixed. It serves to exaggerate changes in the force of gravity by increasing or decreasing it.

The effect on an expanding universe is that it may expand more rapidly than its gravitational content (total mass/energy), in combination with other adjusting factors, would seem to allow.

Gravitational forces would need to have instant influence, essentially move at infinite speed, in order to avoid the increase or decrease of the Doppler effect.

In an expanding universe, the Doppler effect on gravity has the appearance of an expansive force.

For an observer at rest in space, the amount of the Doppler effect on a body's gravity would exceed the real change which the body receives from gravitational migration, just as it does from brightness.

Only for an observer traveling with that body will the Doppler effect cancel itself.

For an observer in any other absolute motion it has an effect which is based on the relative speed between the observer and source.

Earth's Movement

Directional energy shift, which I've described here, is not new or unproven. It was discovered 300 years ago by James Bradley (born in 1693) and it's applied even today by astronomers. It's known as "aberration".

Bradley's discovery was that the locations of all stars appear to shift by 20.5 arc seconds (or about 0.006°) throughout the year. This is a shift of the light we receive from them due to our movement through space. Applying our formula for shift, it tells us that our speed moving through space is about 29,800 m/s (sin 0.006 = 0.0001c = 29,800 m/s). This is the speed of Earth's orbit around our sun.

This is a version of the cosmic speedometer and it tells us not only that Earth is orbiting the sun, but it is not moving significantly otherwise. The reason is because energy shift is based on the absolute speed of the observer moving through space. So it affects the view of all things equally (the stars, distant galaxies, your hand in front of your face).

At this point we have two different answers to the question of what Earth's absolute movement is through space.

- The speed of light (applied through a time dilation experiment by Hafele and Keating) indicates that Earth only rotates in space.

- The direction of light (according to directional energy shift) indicates that Earth also orbits the sun.

I believe there will eventually be an explanation found for this anomaly, but for now will proceed on the basis that Earth's movement is minimal for purposes of evaluating the universe.

Final Thoughts on Absolute Movement

When we consider that movement is absolute, that objects do have a true velocity, not just relative to each other, we can explore the consequences of direction.

Since energy moves independently, and therefore has an absolute direction which is not based on other objects, we're able to use that direction to determine absolute movement of other objects. Consequently, the independence of the direction of energy can also affect the way our world works.

Chapter 10 - Our Absolute Universe

When we take what we've learned from Einstein's relativity, the dilation effects resulting from movement and gravity, and combine them with the absoluteness of movement through space and time, and the directional effects it brings us, we have a new way of evaluating our universe.

The dilation effects of distance and time, and therefore energy, along with the effects of energy shift within a moving system makes calculating speed and distance more complicated. It also allows for deception as the brightness and color shift we see can be caused by more than one scenario.

Deceptive Red-Shift

An example of a deceptive red-shift is how a star moving perpendicular to an observer at rest at a high speed may be red-shifted by dilation and appear the same as another star moving away from the same observer, but being red-shifted by the Doppler effect.

A 2% red-shift may be the result of a star moving away from the observer at 2% of the speed of light. Or, it could be caused by a star 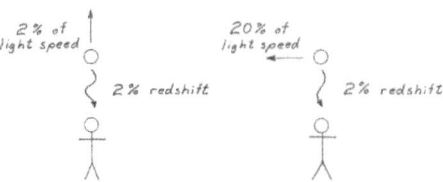 moving at 20% of light speed sideways and emitting a 2% lower energy level (lower frequency) due to it experiencing dilation (living a dilated, or slow life).

Deceptive Dimness

Dilation effects on a traveling star can alter its brightness.

The faster the star is moving, the less energy it will emit, because of its dilated existence, causing it to actually be dimmer.

99

The degree of this can be determined by the standard dilation formula.

If the star is moving away from the observer, energy within it will migrate to its backside causing it to be more active there and brighter by an amount related to its speed as a portion of light

0.2c
2% dilation
9.9% brighter by migration
20% dimmer by Doppler

speed. The Doppler effect will decrease the rate of photons received by the observer, though. And this, too, is related to the speed of the star moving away as a portion of light speed, more than canceling the effect of migration and leaving dilation and a partial Doppler effect to be observed.

If the star is moving toward the observer, energy within it will migrate to its backside, away from the observer, causing its front to be less active and therefore emit less energy

0.2c
2% dimmer by dilation
9.9% dimmer by migration
20% brighter by Doppler

and be dimmer to a degree related to the star's speed as a portion of light speed. The Doppler effect will enhance this light by a greater amount, making the effect unnoticeable. This leaves us with a partial brightening by Doppler.

Regardless of the direction the star is moving, its speed will make it dim and appear further away than it is. So, to proceed in evaluating our universe we'll have to make some assumptions about it.

I'll create a set of hypothetical data based on the popular belief that it's expanding at a constant rate, using 73.04 km/sec per megaparsec (3.26 million light years) of distance, since this is a popularly accepted rate and in line with many other recent estimates. I'll also assume that we're viewing the universe from Earth being nearly at rest relative to space, since we appear to be.

We can, then, deconstruct the hypothetical data and apply absolute relativity to see what it tells us about the true distances and speeds of galaxies.

Hypothetical Universe

Imagine we look through a telescope and see 13 galaxies. They appear to be spread apart by 1 billion light years each in distance from us. There's a 14th galaxy, but we can't see it because it's traveling at nearly the speed of light, so it cannot emit light. In addition, each has a star of the same absolute brightness, so we're able to judge their distance. It's this star which we'll be using for calculations, and each one has a brightness of 1 when viewed from 10 parsecs distance, which is 32.6 light years. I'll also assume that they are all the same mass as our sun, 1.97×10^{30} kg.

Applying the inverse square law:

$$Brightness\ Factor = \frac{1}{\left(\dfrac{d}{32.6\ light\ years}\right)^2}$$

We know what the brightness of each galaxy appears to be. And applying Doppler's formula for red-shift, we know the amount of red-shift we see.

$$\Delta f / f = v/c$$

Our data for this hypothetical universe then, will look like the chart on the following page.

This data, calculated by popular methods, would appear to be of a universe expanding at a constant rate of 73.04 km/sec per 3.26 million light years.

Galaxy	Perceived Distance In Billions Of Light Years	Observed Brightness	Speed According To 73.04 km/sec Rate	Observed Red-Shift
1	1	1.0627×10^{-15}	0.0747	0.0747
2	2	2.6569×10^{-16}	0.1495	0.1495
3	3	1.1808×10^{-16}	0.2242	0.2242
4	4	6.6423×10^{-17}	0.2989	0.2989
5	5	4.2510×10^{-17}	0.3737	0.3737
6	6	2.9521×10^{-17}	0.4484	0.4484
7	7	2.1689×10^{-17}	0.5231	0.5231
8	8	1.6606×10^{-17}	0.5979	0.5979
9	9	1.3120×10^{-17}	0.6726	0.6726
10	10	1.0628×10^{-17}	0.7473	0.7473
11	11	8.7831×10^{-18}	0.8221	0.8221
12	12	7.3803×10^{-18}	0.8968	0.8968
13	13	6.2885×10^{-18}	0.9716	0.9716
14	13.38	5.9364×10^{-18}	1	1

Using the observed brightness and observed red-shift, we can recalculate for true distance and speed according to absolute relativity. We'll need to use new formulas, so let's look at those.

Calculating Speed

Doppler's formula accounts for the effect of shortening wavelengths of a light source and observer moving toward each other regardless of which is doing the moving.

$$\frac{\Delta\lambda}{\lambda} = v/c$$

λ = wavelength

When moving away from each other, frequency should be used, since frequency and wavelength are inversely proportional and the formula needs to provide a 100% (or 1.0) reduction when the speed of light is reached.

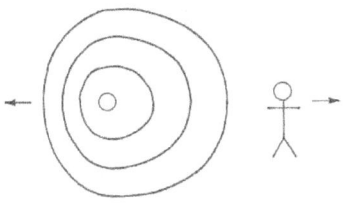

$$\frac{\Delta f}{f} = v/c$$

$$f = \text{frequency}$$

Again, this is regardless of which is moving, because the Doppler effect is an increase or decrease of wavelength or frequency between the source and observer only, regardless of absolute movement.

Einstein recognized that the observer's motion would additionally affect the energy he receives (or experiences). And, in 1905, he wrote his relativistic Doppler formula of:

$$Frequency\ Received = Original\ Frequency \cdot \sqrt{\frac{1 - v/c}{1 + v/c}}$$

This formula accounts for how the energy the observer receives is changed by the Doppler effect, and additionally by the dilation of the observer's own movement. This works for a light source which is stationary when all of the speed is on the part of the observer. (It seems he acknowledged absolute movement here.)

Einstein, and most other people in 1905, believed the universe was static. It wasn't until 1912 that Slipher was finding galaxies to be receding and in 1917 that Willem de Sitter wrote about an expanding universe.

In a situation in which the observer is at rest and the source is in motion, I find the formula to be:

$$Color\ Shift\ =\ 1\ -\ \left(1 - \frac{v}{c}\right)\sqrt{1 - \frac{v^2}{c^2}}$$

This accounts for the Doppler effect and also the reduced energy level of the source due to its movement. It accounts for the source emitting longer wavelength, lower frequency light because of its reduced experience of distance and time. Energy shift and migration have no effect on color.

Using the color shift formula above, we can create the table shown on the following page.

Speed	Colorshift	Speed	Colorshift	Speed	Colorshift
0	0	0.34	0.37932	0.68	0.76537
0.01	0.01005	0.35	0.39111	0.69	0.77562
0.02	0.02020	0.36	0.40291	0.70	0.78576
0.03	0.03044	0.37	0.41471	0.71	0.79578
0.04	0.04077	0.38	0.42651	0.72	0.80569
0.05	0.05119	0.39	0.43830	0.73	0.81547
0.06	0.06169	0.40	0.45009	0.74	0.82512
0.07	0.07228	0.41	0.46187	0.75	0.83464
0.08	0.08295	0.42	0.47364	0.76	0.84402
0.09	0.09369	0.43	0.48539	0.77	0.85325
0.10	0.10451	0.44	0.49712	0.78	0.86233
0.11	0.11540	0.45	0.50883	0.79	0.87125
0.12	0.12636	0.46	0.52052	0.80	0.88000
0.13	0.13738	0.47	0.53219	0.81	0.88858
0.14	0.14847	0.48	0.54382	0.82	0.89703
0.15	0.15962	0.49	0.55542	0.83	0.90518
0.16	0.17082	0.50	0.56699	0.84	0.91319
0.17	0.18208	0.51	0.57851	0.85	0.92098
0.18	0.19339	0.52	0.59000	0.86	0.92856
0.19	0.20475	0.53	0.60144	0.87	0.93590
0.20	0.21616	0.54	0.61283	0.88	0.94300
0.21	0.22762	0.55	0.62418	0.89	0.94984
0.22	0.23911	0.56	0.63546	0.90	0.95641
0.23	0.25064	0.57	0.64669	0.91	0.96269
0.24	0.26221	0.58	0.65786	0.92	0.96865
0.25	0.27382	0.59	0.66896	0.93	0.97427
0.26	0.28545	0.60	0.68000	0.94	0.97953
0.27	0.29711	0.61	0.69096	0.95	0.98439
0.28	0.30880	0.62	0.70185	0.96	0.98800
0.29	0.32051	0.63	0.71266	0.97	0.99271
0.30	0.33224	0.64	0.72339	0.98	0.99602
0.31	0.34399	0.65	0.73402	0.99	0.99859
0.32	0.35576	0.66	0.74457	1.00	1.00000
0.33	0.36753	0.67	0.75502		

With our observed red-shifts, we can look up the actual speeds of those stars and find:

Galaxy	Observed Red-shift	Actual Speed	Galaxy	Observed Red-shift	Actual Speed
1	0.0747	0.07	8	0.5979	0.53
2	0.1495	0.14	9	0.6726	0.59
3	0.2242	0.21	10	0.7473	0.66
4	0.2989	0.27	11	0.8221	0.74
5	0.3737	0.34	12	0.8968	0.82
6	0.4484	0.40	13	0.9716	0.93
7	0.5231	0.46	14	1	1

Calculating Distance

Inverse Square

In order to calculate brightness from distance, we apply the inverse square law in the form of:

$$b_o = \frac{b_a}{(d/32.6)^2}$$

b_o = observed brightness

b_a = absolute brightness

(I use distance in multiples of 32.6 light years because I set all of our stars to a brightness of 1 at that distance. This way we're calculating distance as a multiple of our starting distance, which is 10 parsecs. This formula is customized for that purpose.)

Dilation of Moving Observer

Einstein wrote a modification to the brightness formula to account for an observer being in motion and the light source being at rest.

$$(Observed\ Brightness)^2 = (Original\ Brightness)^2 \cdot \frac{1 - v/c}{1 + v/c}$$

This adjusts for the observer in motion measuring brightness as greater than it really is because his dilation of distance and time make energy seem enhanced. His dilated experience of distance and time creates an enhanced experience of brightness, or amplitude, of what he receives. (He receives the same number of photons, but in a reduced amount of time.) This formula does not include, but works in combination with, the inverse square law.

Dilation of Moving Source

In order to account for an observer at rest measuring a diminished brightness coming from a source in motion, I suggest the standard dilation formula be combined with the inverse square law.

$$A_{v\,dilation} = \sqrt{1 - v^2/c^2}$$

Einstein's formula to adjust for movement of the observer may be left in, since an observer at rest will result in a factor of 1 and have no effect on the result.

Migration

I suggest energy migration be considered as part of the equation. It is an increase based on v/c and it serves to increase the brightness for a source which is moving directly away from the observer. So we can increase our brightness by multiplying by

$$A_{migration} = 1 + \frac{\sum_{\theta=-90}^{\theta=+90} 2\pi \cos\theta \cdot \sin\left[tan^{-1}\left(\tan\theta + \frac{v}{c}\right)\right]}{\sum_{\theta=-90}^{\theta=+90} 2\pi \cos\theta}$$

Doppler

We also saw how the Doppler effect changes the received brightness because, just like with frequency of waves (color, or energy level), it adjusts the frequency of photons (how frequently they are received, brightness). So, we'll multiply by this factor.

$$A_{Doppler} = 1 - v/c$$

107

Dilation Effect Due To Gravity

This effect, while it only becomes significant at the highest speeds, should be included because it will affect how we interpret data from our most distant stars.

$$A_{g\ dilation} = 1 + \frac{Gm}{Gm + c^2 r \sqrt{1 - v^2/c^2}}$$

(I've left out dilation from the gravity of resting mass since that was already accounted for in our absolute brightness. This formula only gives dilation from kinetic energy of the star's travel.)

There is one more consideration, and that is the shutting down of nuclear fusion within stars as they reach high speeds. I don't have a reduction factor to include in our formula, but suspect it may significantly reduce distance measurements. I'll return to this topic later.

We have 6 factors which can adjust the brightness a star appears to be, from its absolute brightness. I'll leave out dilation of a moving observer, since I'm assuming we are at rest in space.

The others may be combined in this way:

$$b_o = \frac{b_a}{(\frac{d}{32.6})^2} (A_{v\ dilation})(A_{migration})(A_{Doppler})(A_{g\ dilation})$$

To solve for distance, we may arrange it as:

$$d = 32.6 \sqrt{\frac{b_a}{b_o} (A_{v\ dilation})(A_{migration})(A_{Doppler})(A_{g\ dilation})}$$

Since we've used red-shift to determine the absolute speeds of stars and know their observed brightness, we can now find our absolute distance.

We can see that the distant galaxies aren't as distant as they seem to be when using only the inverse square law. And, we can calculate the rate of expansion based on each of our galaxies to be:

Galaxy	Billions of Light Years	Galaxy	Billions of Light Years
1	0.9799	8	5.6601
2	1.9091	9	5.8654
3	2.7703	10	5.7912
4	3.5712	11	5.3388
5	4.2554	12	4.5232
6	4.8659	13	2.4789
7	5.3625	14	Indeterminate

Galaxy	H_o	Galaxy	H_o
1	69.82	8	91.51
2	71.67	9	98.31
3	74.08	10	111.38
4	73.89	11	135.46
5	78.09	12	177.18
6	80.34	13	366.67
7	83.84	14	Indeterminate

A changing rate like this could indicate that expansion is slowing with time, since light coming from further away is also coming from equally far back in time. But it could also represent curved paths of galaxies.

Reddening Of Stars

Stars rely on their internal movement to create interactions which fuse atomic nuclei and release energy, light. It is the vibrational energy of heat and it's caused by gravity compressing them.

Some of that movement, even though distances may be short, is near light speed.

As Max Planck discovered with heat energy, the speeds of

particles in a radiating body are of the full range of possibilities. It's the mean speed (the peak of the curve), which equates to mean temperature, that we credit a star with.

Particles in a star at rest in space, then, are free to move at any speed from zero to near light speed, and they do.

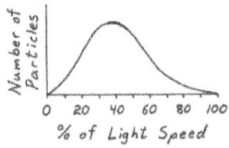

But with the cosmic speed limit of light speed, if the star is to be moving through space, those particles become restricted to a range which is light speed minus the star's speed in the direction of its movement.

If the star is moving at 10% of light speed, then the fastest its particles may move in that direction (in their vibrating) is 90% of light speed, so as to not have an absolute speed greater than light. At a travel speed of 60% of light speed, its particles are limited to vibrating at 40% of light speed in that direction.

And that slowdown in one direction translates into a slowdown of vibrational speed in all directions, as vibrational inertia is lost (transferred to other particles). Any energy the particles are absorbing may still accelerate them as they're able, but will be suppressed by the speed limit each time they move in that direction. This will cause a cooling of the star in proportion to its absolute speed through space, in accordance with its temperature curve.

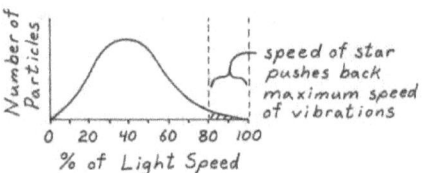

As speed increases, the maximum speed of particles gets reduced. The effect becomes dramatic as it begins to eat away substantial portions of the curve.

The lost inertia must be conserved, and its options are to escape as energy, which is extremely difficult from the core of a star, or to pass on its energy to slower moving particles. This alters the shape of the whole curve. And a change in color occurs, a reddening as the average speed (temperature) is reduced well before the high-speed end of the curve is substantially eroded. The smallest portion at the high-speed end of the curve immediately finds a place in the

110

slower part of the curve. As it does, that slower portion is amplified. This effect is exaggerated by each high-speed particle having so much energy that they likely pass it on to, not one, but multiple lower energy particles.

This is different from a color shift of light received from a star. This reddening is a change in the absolute color of the star occurring prior to any color shift which may be measured by shifted spectral lines.

The color shifts which alter the positions of spectral lines occur as the light passes through the atmosphere of the star. The spectral lines originate in the star's atmosphere. The reddening from 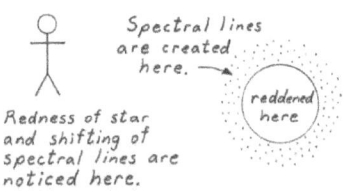 the star cooling happens in the star itself before the spectral lines are created.

This is another reason why stars of more distant galaxies appear redder than their spectral classes indicate they should be. Spectral classes are designated primarily because of the elements contained in the star, which determine the temperature the star should burn at (the speed the particles should vibrate at).

Dimming of Stars

As stars are forced to cool, or their temperature is limited, because of their high absolute speed, they also dim. The total energy emitted from them becomes less. The number of photons emitted are fewer because the rate of collisions goes down. And it's the collisions that succeed in fusing particles together which release the high energy photons, the bulk of the star's energy.

When a collision finally breaks the electromagnetic barrier (the force preventing positively charged particles from getting too close to each other), they enter the range of the strong force. The particles then are bonded by the strong force, and a gamma ray (a high energy photon) is released. It's this high energy photon which gets absorbed and re-emitted on its way to the surface of the star to escape as multiple lower energy particles.

Fusion requires powerful high speed impacts. And when the particles can't reach that speed, the probability of fusion drops dramatically. Without production of elements, the gamma rays aren't released, and a star's main source of energy emission is diminished.

Stars in motion then, struggle to be stars more, the faster they travel. It's the slow moving stars which produce the elements and power the universe.

Considering only the inverse square law applied to brightness, stars could seem to be great distances away. Very great distances could be calculated in this way if the observer could detect a low enough brightness. But, absolute relativity tells us a star will dim entirely before it reaches light speed. That means any stars far enough away to be traveling near enough to light speed, according to expansion, are emitting no light to be seen by anyone. And the very dim ones we can detect are not as far away as we might believe, but are probably within that distance where their speed is still slow enough to emit light.

The Horizon Problem

As we gained the ability to see dimmer light sources deep in space with more sensitive telescopes, we also learned to calculate the age of the universe better. But the age of the universe and distance of stars are not compatible. Finding dimmer stars has led us to believe we've been finding the universe to be bigger. Too big for the consistency which is revealed to us by the background radiation coming to us from every direction, the microwaves from the early universe which continue to travel today.

Simply, if the universe is only about 13 and a half billion years old and it was once all together sharing temperature equilibrium and nothing can travel faster than light, then why do we see galaxies which are so dim that they appear to be 15 billion, 20 billion, or more light years away? Some believe we can see galaxies which are 50 billion light years away. How did they get there? A period of inflation, expansion faster than light speed, has been proposed using quantum effects to get around the speed limit. But, I suggest recalculating distances with a consideration of the effects of absolute movement. I believe it will be found that the visible horizon (the furthest stars we can see) is much less than is currently believed and there may not be a need for a period of inflation to explain how galaxies have reached their distance.

Adjustments To Gravity

Based on the effects absolute relativity has on gravity, we should consider how they change the total gravitational force between bodies.

Let's look at each effect and what the result of it would be in the interaction of 2 independent bodies which have their own unique velocities.

Dilation seems that it cannot influence gravity. As evidence are black holes, which should have a fully dilated existence and yet seem to interact gravitationally with no change.

Energy shift, as it relates to gravity, is equal to the size enlargement or reduction as described in "Visual Distortion". To determine if gravity between 2 moving bodies is enhanced or reduced by their travel speed, each must be calculated separately. For example, looking at our chart, we can see that a body moving at 40% of light speed (0.4c) which is actually 30° behind another body will experience that body's gravitational pull as reduced by 31.5% and in a direction of 44.3°. That other body, if traveling with it, will experience gravity from the first one as being enhanced by 29.0% and from a direction of 10.1° to its rear.

Energy migration is a unique effect as it is a real change in gravitational force. And, while it cancels between 2 bodies moving the same, 2 bodies moving differently experience their difference.

The Doppler effect is a simple proportional change in the force of gravity based on the relative speed between the bodies involved. And its direction is determined by energy shift.

While we're familiar with kinetic energy (the energy of movement) already, it is absolutely an absolute relativity concept since it is determined based on absolute speed of a body. Any kinetic energy must be based on absolute movement, otherwise it would be different for every observer.

Kinetic energy, being energy, has a mass equivalency. And the amount of that mass equivalency grows in direct proportion to the speed of the object.

If you're curious of how we know it's dependent on absolute movement, consider going back to our thought experiment. You'll see how your movement, to me, would give you extra mass equivalency from kinetic energy, and my movement, to you, would give me extra mass equivalency from kinetic energy.

Dark Matter

These effects could dramatically alter the positions and movements of objects in a traveling system, such as a galaxy. Energy shift predicts how a body in orbit will experience its attractor (the massive object it orbits) as being toward its front, rather than perpendicularly. And, energy shift as well as the Doppler effect will cause it to experience a greater gravitational pull from its attractor than it really has. Bodies in orbit will accelerate.

This may help to explain the anomaly of excessively fast moving outer orbits in galaxies, which is often credited to dark matter. It may also explain what is interpreted as frame dragging.

Chapter 11 – Proof of Movement

In the first chapter I began explaining how we can be sure that movement actually occurs according to the movement theory described in my book *"Out of This World: The Movement Dimension"*. This theory states that objects leave space in order to relocate. The test for this is to observe matter passing through matter.

We can't see this occurring in our daily lives due to our relatively slow speeds and large objects. But it can be observed in particle accelerators which collide (or attempt to collide) small objects at high speeds.

Now that you have an understanding of the effects of absolute relativity, let's look at data from an accelerator and see how it aligns with predictions made by movement theory.

Particle Accelerators

These devices collide particles at the high speeds which would result in a measurable degree of particles passing through each other. And a quick look at their data will reveal that it looks as though it may actually be happening.

The Large Hadron Collider at CERN, in Switzerland, runs 2 streams of protons into each other at 99.9999991% of light speed each. Each stream is 16 microns in diameter and contains 115 billion protons. The result is an average of about 20 collisions. To determine how many particles may be passing through each other, we first need to determine how many collisions they should have. Protons are very small and 16 microns of diameter is quite large in comparison.

A quick preliminary calculation tells us that 115 billion protons, which are about 1.71×10^{-15} m in diameter (or 2.297×10^{-30} m in area) would occupy 2.641×10^{-19} m of area total. And the area available to them, the cross-section of the stream, is 2.01×10^{-10} m.

Assuming the protons are evenly distributed, they only take up 1.314 x 10^{-9}, or 0.000000001314 of the area available. This tells us that each proton has a 1 in 761 million chance of colliding with another proton. And 115 billion protons have a 115 billion in 761 million chance. This means 151 protons should collide with a counterpart.

But only 20 collide. There appears to be something going on.

Proton Distribution & Size

Let's consider additional factors which could affect predictions. These fall into the categories of distribution and effective size. The protons should not be distributed evenly and they should be effectively larger than just their particle size.

The particles are not flowing like a gas or liquid goes through a tube, being propelled by pressure. They are held into their 16 micron wide stream by magnetic force. (The "tube" they are in is much larger than that.)

Being pushed on all sides by magnetism, every particle should want to be in the center. If they were, the stream would be the width of 1 proton and they would all collide when they meet the opposing stream. Instead, they most probably are in the center. And, the 16 micron stream width doesn't define where the protons are, since they may be anywhere in the width of the actual tube. The 16 micron width tells us where they most probably are. It tells us of the bulk of the protons. Not 100% of them, but the great majority.

The16 micron width is a measurement of most of a bell curve type distribution. From this we know a couple things. One is that the far reaches of the distribution, the flared out ends of the curve, are very sparsely populated and very unlikely to collide. Another is that the center, where the highest concentration is, will have a greatly increased chance of collision. Overall, this type of distribution raises our expected collisions significantly.

The effective size of a proton is greater than its particle size.

At a high speed, it is much greater. This is based on several factors.

Protons are all positively charged. When 2 positively charged particles run into each other, they don't get close enough to actually touch. Their resistance (due to their like charge) increases exponentially as they approach each other. They will fuse into a new particle before they touch.

They effectively collide at a distance of about 5.636×10^{-15} m. If a charged particle was colliding with a non-charged particle, such as a neutron, then only physical size would apply. Because we're looking at all protons, we can add this additional distance to the proton's physical size to determine its effective size.

Effects of Absolute Relativity

The other factors I find to be relevant are a result of velocity of the protons and are effects of absolute relativity. For now, let's recalculate using only these classical concepts to see where we're at, to see how much of a phenomenon we have at this point.

The reason for doing this is that absolute relativity, and the effects it predicts which I'll be applying, depends on movement theory (or one such as it) as a model. Here I'm attempting to prove movement theory correct using the effects of absolute relativity for evidence. The two concepts rely on each other. Some people may prefer to dismiss both, and say that the collisions at the Large Hadron Collider at CERN only involve the distribution and size factors we've considered so far.

Considering the physical size of a proton (1.71×10^{-15} m diameter) and increasing it by 2.818×10^{-15} m all around to account for the influence of its charge, each proton has an effective radius of 3.673×10^{-15} m. This makes the area of each proton 4.238×10^{-29} m^2. This enlargement greatly increases the chance of collisions.

If the protons are distributed according to probability, then we can calculate their concentration at distances from the center of the stream of particles. We only need to determine what percentage of the particles lie within the 16 micron width of the stream.

It is impractical to say 100% because probability predicts the distribution to extend indefinitely on all sides. One hundred percent of the protons inside the 16 micron stream would have most all of the protons lined up in a row for one massive collision, or the equivalent of almost 115 billion individual collisions.

Perhaps there would be a handful of protons escaping this event. Not knowing what figure is used here makes this factor speculative and subject to verification. For purposes of these calculations, I'll use 95%. This gives us a varied distribution of protons within the stream, which still looks like a stream of 16 microns, yet allows for plenty of protons to essentially not participate.

Two streams of 115 billion protons each, which are 95% within a 16 micron diameter, in which each proton is effectively 7.346×10^{-15} m wide, should result in 15,634 collisions.

Since experience shows us that only an average of 20 collisions occurs during each of those events, this appears to be a significant discrepancy with no logical explanation. Movement's prediction of particles leaving space in order to travel might seem like a possible reason.

However, since these protons are traveling at 99.9999991% of the speed of light, movement predicts that only 2 collisions should occur. At that high speed, each proton would be 99.9866% in wave form and only 00.0134% in the form of a particle capable of colliding.

Absolute relativity acknowledges that every object has one absolute speed and direction in relation to space. This results in effects for both matter and energy.

Protons, being charged particles, communicate their charge by photons. The way protons are able to push away from each other is by their photon exchange. Photons, being energy, are subject to effective increases or decreases due to the movement of the objects which exchange them. These result from distance and time dilation of the traveling objects as well as the Doppler effect, directional energy shift, and migration.

All protons in the accelerator scenario will experience the same amount of dilation. Every proton would emit reduced energy according to its dilated existence and experience each other's reduced energy as being enhanced back up to a normal level, so the effect will cancel. And dilation will cause no noticeable change in their energy exchange.

Migration will not affect the individual relationships between 2 single protons, since each proton is not a system. It can become evident within the streams of protons, but because the particles are still quite distant from each other (as compared to the particles making up a piece of metal) I will not be considering its effects. If it were to show itself, it would be as an increase in the effective size of the protons to the rear of the stream and a decrease to the ones in front.

Directional Shift

Directional energy shift is a unique and key factor affecting the size of high speed charged particles. At the speed of particle accelerators, very near light speed, the direction of the photons carrying the electric charge of each proton is most dramatically changed.

All energy received by these traveling protons is received at a much more forward angle than if they were not traveling. As two of these protons attempt to pass by each other, the directional shift has its greatest effect just before they reach a relationship of perpendicular to travel. When they are 10° in front of each other, they will experience the energy they receive from the other as if it was coming from 49.6° in front of them. And this energy will be concentrated by 227%. This will extend the charge area of each proton by 227% in that 10° forward direction giving each proton a 223% wider charge area. (cos 10° x 227% = 223%) Their 2.818 x 10^{-15} m electrical barrier becomes 6.284 x 10^{-15} m.

Doppler Effect

The Doppler effect increases the concentration of energy a traveling object runs into by the amount of the speed of the traveling object, as a portion of light speed. If the protons were running into energy at half of light speed, they would run into 50% more energy per second than what is actually coming at them, 50% more photons. In this case, protons are running into each other's energy at 99.9999991% of light speed. This doubles the concentration (raises it by 99.9999991%). So, the electrical reach is raised by nearly 100% to 1.257×10^{-14} m.

Our protons, because of their travel speed, are effectively 1.342×10^{-14} m in radius (0.855×10^{-15} m + 1.257×10^{-14} m). The area each occupies is now 5.658×10^{-28} m^2, much more than our classical estimate of 4.238×10^{-29} m^2.

With an effective proton area of 5.658×10^{-28} m^2, and with 95% of our probability distribution within the 16 micron streams, we can predict that 208,727 collisions will occur.

Summary

Now, reconsidering movement's prediction that each of these particles are only 0.0134% likely to exist in particle form at any one moment, only 28 of these collisions should occur. And 208,699 of these protons should pass through their oncoming proton.

While the figures I used here may have various degrees of accuracy, I believe the calculations as a whole provide strong support that not only are these protons passing through each other, but they are leaving space to do so in accordance with the model I proposed for how movement occurs and in support of the effects of absolute relativity.

Chapter 12 - Conclusion

Relativity is the underlying logical structure which we use to learn about our world. It describes the connectedness of all things. It's what tells me that I'm here, and that you're here too. While we experience the world differently, it's still the same world.

Relativity is what makes our world knowable.

It may be imagined as a vast network of individual relationships among all things which appear different to each observer. But that network is something too. The existence of individual relationships demands the existence of their absolute relationship. One can't have reliable interactions with all of the parts without there arising a whole logical system. It's like the forest and its trees. The whole must exist because of its parts. And we can be sure of the interconnectedness of all things because of the unlimited reach of gravitational and electromagnetic forces.

It's been popular to describe our world as being dependent on the observer, but even then there is some agreement as to an absolute, or privileged reference frame. We generally reference the Earth, or the Earth as if it weren't rotating.

This is because we're not just describing appearances, we're describing events which are related. And it's the events which prove our laws and make our world knowable.

If events were not absolute, and they were to vary for each of us, then the laws would, too. There would not be a consistency to nature, and it would not be knowable.

We wouldn't think to have a principle of relativity if it wasn't preceded by consistency. From consistency comes our world being knowable and proof of our relationship to a common reality.

I believe the ideas I've put forth here are securely founded in established science as well as our own experience. My intent was to demonstrate how relativity may be extended to help us understand our world further.

We know a great deal about our world through many discovered concepts. My hope is that existing concepts will be taken to their fullest application through the absoluteness which unites our whole universe.

About the Author

Timothy Michaels is an artist and writer. Being a technically minded, "cerebral" thinker, he enjoys creating realistic art as well as exploring the world around him through physics. This led him to an explanation of color relationships using physics, which resulted in the development of his Color Calculator. Though he has no formal education in physics, Mr. Michaels has an exceptional gift for logic and an insight into the workings of the universe.

His scientific ideas rely on established theories with evidence, and are explained in a simple manner for general readership. Mr. Michaels provides new ways of understanding these theories, and his ideas go beyond current science to solve problems and make new predictions. In doing so they provide the concepts and formulas necessary to advance science. Students of physics, astronomy, cosmology and other fields will find his explanations enlightening and his new ideas worth investigating.

Other Books by Timothy Michaels

"Absolute Relativity: How Newton and Einstein Agree"

"Out of This World: The Movement Dimension"

"The Physics of Color Harmony"

"How We See Art"

"Walking the Cards: A Unique Drawing Method"

His artwork may be viewed at: www.tmsartgallery.com

Readers are invited to comment by sending an email to: 101timsplace@gmail.com

Bibliography

Clegg, Brian. *The Universe Inside You*, Icon Books Ltd., 2012

Ferris, Timothy. *Coming of Age in the Milky Way*, Anchor Books, 1998

Fleming, Thomas A. (editor). *Stars*, Cognella, 2011

Folger, Tim. *Voyagers to the Stars*, Scientific American, July 2022

Glover, Linda K. with Chaikin, Andrew. Daniels, Patricia S. Gianpoulos, Andrea. Malay, Jonathan T. *National Geographic Encyclopedia of Space*, National Geographic, 2005

Gribbin, John. *Einstein's Masterwork*, Pegasus Books Ltd, 2016

Hawking, Stephen W. *A Brief History of Time: From the Big Bang to Black Holes*, Bantam Books, 1988

Impey, Chris. *Einstein's Monsters: The Life and Times of Black Holes*, W.W. Norton & Company, Inc., 2018

Lorentz, H.A. Einstein, A. Minkowski, H. Weyl, H. *The Principle of Relativity: A Collection of Original Papers on the Special and General Theory of Relativity*, Dover Publications, Inc., 1952

Menzel, Donald H. (editor). *Fundamental Formulas of Physics*, Dover Publications, Inc., 1960

Nassau, Kurt. *The Physics and Chemistry of Color*, John Wiley & Sons, Inc., 1983

Panek, Richard. *A Cosmic Crisis*, Scientific American, March, 2020

Wolfson, Richard. *Simply Einstein: Relativity Demystified*, W.W. Norton & Company, Ltd., 2003

www.ingramcontent.com/pod-product-compliance
Lightning Source LLC
Chambersburg PA
CBHW072308290526
45794CB00002B/568